THE HAM RADIO BIBLE

The Definitive Guide to Mastering Ham Radio in No Time, with In-Depth Explanations and Practice Questions to Get Ready for the Technician Class Amateur Radio Exam

Frank Webb

© Copyright 2024 by Frank Webb. All Rights Reserved.

The publication is sold with the idea that the publisher is not required to render accounting, officially permitted or otherwise qualified services. This document is geared towards providing exact and reliable information concerning the topic and issue covered. If advice is necessary, legal or professional, a practiced individual in the profession should be ordered.

- From a Declaration of Principles which was accepted and approved equally by a Committee of the American Bar Association and a Committee of Publishers and Associations.

In no way is it legal to reproduce, duplicate, or transmit any part of this document in either electronic means or printed format. Recording of this publication is strictly prohibited, and any storage of this document is not allowed unless with written permission from the publisher—all rights reserved.

The information provided herein is stated to be truthful and consistent. Any liability, in terms of inattention or otherwise, by any usage or abuse of any policies, processes, or directions contained within is the sole and utter responsibility of the recipient reader. Under no circumstances will any legal responsibility or blame be held against the publisher for any reparation, damages, or monetary loss due to the information herein, either directly or indirectly.

Respective authors own all copyrights not held by the publisher.

The information herein is offered for informational purposes solely and is universal as so. The presentation of the information is without a contract or any guarantee assurance.

The trademarks that are used are without any consent, and the publication of the trademark is without permission or backing by the trademark owner. All trademarks and brands within this book are for clarifying purposes only and are owned by the owners themselves, not affiliated with this document

Go to the last page to download the bonus questions!

Index

- Introduction 6
- Welcome to Ham Radio! 6
- The Fascination of Ham Radio 6
- The Importance of Licensing 7
- Overview of Ham Radio History 7
- Getting Started 8
- Basic Concepts and Terminology 9
- Equipment You Will Need 10
- Setting Up Your First Station 11
- CHAPTER 1 – Licensing and Regulations 11
- Understanding the Licensing System 11
- Different Classes of Licenses 12
- The Licensing Process 13
- Rules and Regulations 14
- FCC Rules and Regulations 15
- Operating Privileges and Restrictions 16
- International Regulations 16
- Call Signs 17
- Structure and Allocation 17
- Vanity Call Signs 18
- CHAPTER 2 – Operating Practices 18
- Basic Operating Procedures 18
- Communication Protocols 20
- Q-Signals and Common Abbreviations 21
- Advanced Operating Techniques 22
- Digital Modes 23
- Satellite Communication 25
- Emergency Communications 26
- Preparedness and Protocols 27
- Participating in Emergency Nets 28

CHAPTER 3 – Radio Technology Fundamentals 30
Basic Electronics and Components 30
Ohm's Law and Basic Circuits 31
Components: Resistors, Capacitors, Inductors 32
Understanding Radio Waves 33
Properties of Radio Waves 34
Frequency and Wavelength 34
Propagation Basics 35
CHAPTER 4 – Circuitry and Design 36
Circuit Analysis 36
AC and DC Circuits 38
Impedance and Reactance 41
Building and Troubleshooting Circuits 43
Schematic Reading 46
Practical Circuit Design 49
CHAPTER 5 – Licensing and Regulations Transmission Lines and Antennas 53
Transmission Line Basics 53
Types of Transmission Lines 56
Impedance Matching 59
Antenna Theory and Design 62
Types of Antennas 66
Antenna Installation and Safety 71
Grounding and Lightning Protection 74
CHAPTER 6 – Radio Equipment 77
Transceivers and Receivers 77
Types and Functions 77
Selecting the Right Equipment 79
Amplifiers and Power Supplies 80
Understanding Amplifier Types 81
Practical Example: Setting Up a Base Station Transceiver 83
CHAPTER 7 – Digital Modes and Software 84
Introduction to Digital Modes 84
Popular Digital Modes 85

Software for Ham Radio 86

Software for Digital Modes 87

CHAPTER 8 – Propagation and Space Weather 90

Understanding Propagation 90

Ionospheric Layers 92

Day and Night Propagation 94

Space Weather and its Effects 96

Solar Activity 98

Predicting Propagation Conditions 100

CHAPTER 9 – Building and Mantaining Your Station 103

Station Design and Layout 103

Ergonomics and Efficiency 104

Safety Considerations 106

Maintenance and Upkeep 109

Routine Checks and Troubleshooting 111

Upgrading Equipment 113

CHAPTER 10 – Operating Awards and Contests 115

Introduction to Awards and Contests 115

Popular Awards: DXCC, WAS, WAC 116

Major Contests: Field Day, Sweepstakes 117

Strategies for Success 118

Contesting Techniques 119

Award Hunting Tips 120

CHAPTER 11 – Advanced Topics 122

Homebrewing Equipment 122

Building Your Own Gear 124

Kits and Resources 126

Software-Defined Radio (SDR) 127

Advanced Data Modes 129

CHAPTER 12 – Community and Resources 131

Joining Ham Radio Clubs 131

Conclusion and BONUS 133

Introduction

Welcome to Ham Radio!

Welcome to the fascinating world of ham radio, a hobby that has captivated millions around the globe. Ham radio, also known as amateur radio, offers a unique blend of technical challenge and community engagement, making it a hobby like no other. Whether you are interested in communicating with people across the world, learning about electronics, or providing emergency communication services, ham radio has something to offer.

As you embark on this journey, you will discover the rich history, diverse modes of communication, and the vibrant community that makes ham radio so special. This book is designed to guide you through the essential knowledge and skills needed to become a proficient ham radio operator. We will cover everything from licensing and regulations to operating practices, technical fundamentals, and building your own station. Whether you're a beginner or looking to deepen your expertise, this comprehensive guide will help you navigate the exciting world of ham radio.

In this guide you'll notice that some concepts are repeated. This intentional choice is made to ensure better assimilation of the information. Repetition helps reinforce understanding and makes key ideas more familiar, thereby facilitating the learning process and practical application of the knowledge gained.

The Fascination of Ham Radio

Ham radio has an enduring appeal that continues to attract enthusiasts of all ages. The allure lies in its versatility and the endless possibilities it offers. For some, the fascination begins with the ability to connect with people from different cultures and countries, often from the comfort of their own home. Imagine the thrill of speaking to someone on the other side of the world using just your radio equipment and a set of antennas.

Others are drawn to the technical aspects of the hobby. Ham radio provides a hands-on way to learn about electronics, radio wave propagation, and the intricacies of building and maintaining radio equipment. The satisfaction of constructing your own radio or antenna and then using it to make contact is unparalleled.

Emergency communication is another compelling aspect of ham radio. When traditional communication networks fail, ham radio operators step in to provide critical links. This role in disaster response and community service underscores the importance and utility of ham radio in our modern world.

The social aspect of ham radio should not be overlooked. Joining a ham radio club or participating in contests and events offers opportunities to meet like-minded individuals, share knowledge, and develop lifelong friendships. The camaraderie and shared passion within the ham radio community are integral parts of the hobby's charm.

In essence, ham radio is more than just a hobby; it is a gateway to a world of exploration, learning, and connection. Whether you are driven by a curiosity for technology, a desire to communicate, or a commitment to public service, ham radio offers a rich and rewarding experience.

The Importance of Licensing

Obtaining a ham radio license is a crucial step in becoming an amateur radio operator. Licensing ensures that operators have the necessary knowledge to use the radio spectrum responsibly and effectively. It also helps regulate the airwaves to prevent interference and maintain the integrity of communications.

The licensing process involves passing an examination that tests your understanding of radio theory, operating practices, and regulations. While this might seem daunting at first, the process is designed to be accessible and educational. Studying for your license will equip you with a foundational knowledge of electronics, radio frequency propagation, and the rules governing amateur radio.

There are different classes of licenses, each offering varying levels of privileges. In the United States, the Federal Communications Commission (FCC) oversees the licensing process. The entry-level Technician class license is relatively straightforward to obtain and allows you to access many VHF and UHF frequencies. As you advance, the General and Amateur Extra class licenses open up more frequencies and operating privileges.

Licensing serves several important purposes. It helps ensure that operators are knowledgeable about safe and effective operating practices. This is particularly important when it comes to issues like avoiding interference with other communications, adhering to power limits, and using the correct frequencies. Additionally, licensed operators are better equipped to provide emergency communications and contribute to public service events.

Moreover, obtaining a license is a source of pride and accomplishment. It marks your entry into a community of skilled and knowledgeable individuals who share a passion for radio communication. The knowledge and skills you gain during the licensing process will serve you well throughout your ham radio journey.

In summary, licensing is a vital aspect of ham radio that promotes responsible use, enhances technical knowledge, and fosters a sense of community among operators. It is an essential step towards unlocking the full potential of what ham radio has to offer.

Overview of Ham Radio History

The history of ham radio is rich and varied, tracing back to the early days of wireless communication. The journey began in the late 19th and early 20th centuries with the pioneering work of scientists like Heinrich Hertz, Guglielmo Marconi, and Nikola Tesla, who laid the foundations for modern radio technology.

Guglielmo Marconi's successful demonstration of wireless telegraphy in the late 1890s marked a significant milestone. His ability to send signals across the Atlantic Ocean in 1901 showcased the potential of radio waves for long-distance communication. This breakthrough spurred further experimentation and innovation, leading to the development of more sophisticated radio equipment and techniques.

The early 20th century saw the emergence of amateur radio enthusiasts who experimented with homemade equipment and established some of the first radio communication networks. These early "hams" played a crucial role in advancing radio technology and demonstrating its practical applications. They experimented with different frequencies, modes of transmission, and antenna designs, contributing to a growing body of knowledge about radio communication.

The formation of the American Radio Relay League (ARRL) in 1914 was a pivotal moment in ham radio history. Founded by Hiram Percy Maxim, the ARRL provided a formal organization for amateur radio operators and helped coordinate the relay of messages across long distances. The ARRL's efforts were instrumental in gaining recognition and support for amateur radio from government authorities.

Ham radio also played a vital role during World Wars I and II. Many amateur operators were enlisted to serve as radio operators, using their skills to support military communications. The post-war period saw a resurgence in amateur radio activity, with advances in technology making equipment more accessible and affordable.

The introduction of licensing requirements and regulations in the mid-20th century helped formalize the hobby and ensure that operators were knowledgeable and responsible. Technological advancements, such as the development of transistors and the advent of digital communication modes, further expanded the capabilities of ham radio.

Today, ham radio continues to evolve, embracing new technologies like software-defined radio (SDR) and digital communication modes. The hobby remains a vibrant and dynamic field, attracting enthusiasts from all walks of life who share a passion for radio communication.

In summary, the history of ham radio is a testament to human ingenuity and the spirit of experimentation. From its humble beginnings to its current status as a global hobby, ham radio has continually adapted and thrived, contributing to the advancement of communication technology and fostering a worldwide community of operators.

Getting Started

Embarking on your ham radio journey is an exciting endeavor that offers numerous opportunities for learning and connection. The first step is to familiarize yourself with the basics and gather the necessary resources to get started.

One of the most important aspects of getting started is finding reliable information and guidance. There are numerous resources available, including books, online forums, and local ham radio clubs. The ARRL (American Radio Relay League) is a valuable resource that offers publications, courses, and support for new and experienced operators alike. Joining a local ham radio club can also provide mentorship, hands-on experience, and a sense of community.

The next step is to study for your ham radio license. The licensing process varies by country, but it generally involves passing an exam that covers basic electronics, operating practices, and regulations. In the United States, there are three levels of licenses: Technician, General, and Amateur Extra. The Technician license is the entry-level license and is a great starting point for beginners. There are many study guides and online resources available to help you prepare for the exam.

Once you have your license, you can start thinking about the equipment you will need. At a minimum, you will need a transceiver, an antenna, and a power supply. Many beginners start with a handheld transceiver (HT), also known as a "handy-talkie," which is portable and relatively inexpensive. As you gain experience and knowledge, you may want to invest in more advanced equipment, such as a base station transceiver and a variety of antennas.

Setting up your first station is an exciting milestone. Choose a location in your home where you can set up your equipment comfortably. This could be a dedicated radio room or a corner of a room with enough space for your transceiver, power supply, and other accessories. Ensure that your antenna is installed safely and securely, and that your equipment is properly grounded to prevent electrical hazards.

As you start operating, it's important to learn and follow proper operating procedures and etiquette. This includes understanding how to make a call, how to log your contacts, and how to use Q-signals and other abbreviations. Listening to experienced operators and participating in local nets (regularly scheduled on-air meetings) can help you develop good habits and improve your skills.

Finally, don't be afraid to ask for help and advice. The ham radio community is known for being friendly and supportive. Whether you have technical questions, need help troubleshooting, or want to learn about new aspects of the hobby, there are many experienced operators who are willing to share their knowledge and assist you.

In summary, getting started with ham radio involves finding reliable information, studying for your license, gathering equipment, setting up your station, and learning proper operating procedures. With dedication and enthusiasm, you will soon be making contacts and enjoying the many benefits of this fascinating hobby.

Basic Concepts and Terminology

To navigate the world of ham radio, it is essential to understand some basic concepts and terminology. Here are a few key terms that you will encounter frequently:

- **Frequency:** The number of times a wave oscillates per second, measured in hertz (Hz). In ham radio, frequencies are often measured in kilohertz (kHz), megahertz (MHz), or gigahertz (GHz).

- **Wavelength:** The distance a radio wave travels during one complete cycle of oscillation. Wavelength and frequency are inversely related.

- **Transceiver:** A device that combines a transmitter and a receiver in one unit, used for sending and receiving radio signals.

- **Antenna:** A device that radiates and receives radio waves. Different types of antennas are used for different frequencies and purposes.

- **Modulation:** The process of varying a carrier signal to transmit information. Common types of modulation include amplitude modulation (AM) and frequency modulation (FM).

- **Call Sign:** A unique identifier assigned to a licensed ham radio operator, used for identification during communications.

- **Q-Signals:** A standardized set of codes used to convey common phrases efficiently, such as "QTH" for location and "QRZ" for who is calling.

Understanding these basic terms will help you communicate effectively and navigate the technical aspects of ham radio.

Equipment You Will Need

Starting with the right equipment is crucial for a successful ham radio experience. Here are the essential items you will need:

- **Transceiver:** A transceiver is a combination of a transmitter and a receiver in one unit. For beginners, a handheld transceiver (HT) is a good starting point. As you advance, you may opt for a base station transceiver.

- **Antenna:** The antenna is a critical component that radiates and receives radio signals. Different types of antennas, such as dipole, vertical, and Yagi, are used for various frequencies and applications.

- **Power Supply:** Your transceiver needs a reliable power source. Handheld transceivers usually come with rechargeable batteries, while base stations often require an external power supply.

- **Coaxial Cable:** This cable connects your transceiver to the antenna, ensuring efficient signal transmission.

- **Grounding System:** Proper grounding is essential for safety and optimal performance. A good grounding system protects your equipment from electrical surges and reduces interference.

- **Accessories:** Depending on your interests, you might need additional accessories such as a microphone, headphones, and an SWR meter for tuning your antenna.

Investing in quality equipment and setting it up correctly will enhance your ham radio experience and ensure you get the most out of your hobby.

Setting Up Your First Station

Setting up your first ham radio station is a rewarding experience. Here's a quick guide to get you started:

1. **Choose a Location:** Find a quiet, comfortable space with enough room for your equipment.
2. **Install Your Antenna:** Ensure it's securely mounted and properly aligned for optimal performance.
3. **Connect Your Equipment:** Use coaxial cables to connect your transceiver to the antenna and power supply.
4. **Ground Your Station:** Implement a proper grounding system to protect against electrical surges.
5. **Test Your Setup:** Power on your equipment and perform initial tests to ensure everything is functioning correctly.

With your station set up, you're ready to start exploring the world of ham radio.

CHAPTER 1 – Licensing and Regulations

Understanding the Licensing System

The licensing system for amateur radio operators is designed to ensure that individuals have the necessary knowledge and skills to operate their equipment safely and effectively. This system not only promotes responsible use of the radio spectrum but also helps to minimize interference with other communication services. In the United States, the Federal Communications Commission (FCC) oversees the licensing process, which involves passing an examination that tests an applicant's understanding of radio theory, regulations, and operating practices.

A well-structured licensing system serves multiple purposes. It ensures that radio operators have a foundational understanding of key concepts such as frequency allocation, modulation techniques, and emergency communication procedures. By requiring operators to demonstrate their knowledge through exams, the system helps maintain high standards within the ham radio community.

Examinations for amateur radio licenses are typically administered by Volunteer Examiners (VEs), who are experienced operators authorized by the FCC to conduct testing sessions. These VEs ensure that the examination process is fair and consistent, providing a reliable measure of an applicant's readiness to become a licensed operator.

The licensing system also introduces new operators to the various aspects of amateur radio, encouraging them to explore different modes of communication, participate in public service events, and engage with the global ham radio community. By learning about the rules and best practices, operators can contribute to a more organized and efficient use of the radio spectrum.

Studying for the licensing exam can be a rewarding experience in itself. Many resources are available to help aspiring hams prepare, including books, online courses, and practice exams. The American Radio Relay League (ARRL) offers a comprehensive set of study materials tailored to each class of license, ensuring that candidates have access to the information they need to succeed.

For beginners, the Technician class license is the entry point. It covers the basics of amateur radio, including operating procedures, basic electronics, and VHF/UHF band privileges. With a Technician license, new operators can start communicating on local repeaters, participate in local nets, and experiment with digital modes.

As operators gain experience and knowledge, they can advance to higher license classes, such as General and Amateur Extra. Each successive class grants additional operating privileges, allowing access to more frequency bands and modes of communication. This progression encourages continuous learning and skill development, fostering a deeper understanding of the technical and regulatory aspects of ham radio.

The licensing system also plays a crucial role in ensuring the safety and security of amateur radio operations. Licensed operators are trained to follow proper procedures, avoid interference with other services, and adhere to power limits. This helps maintain the integrity of the radio spectrum and prevents disruptions to critical communication networks.

In addition to national regulations, amateur radio operators must also be aware of international agreements and guidelines. Organizations such as the International Telecommunication Union (ITU) work to harmonize radio regulations across different countries, promoting global cooperation and coordination. By understanding and following these international standards, operators can participate in global communications and contribute to international goodwill.

Overall, the licensing system is a cornerstone of the amateur radio hobby. It ensures that operators are well-prepared, knowledgeable, and responsible, fostering a vibrant and skilled community of radio enthusiasts. By adhering to the rules and continually expanding their knowledge, licensed operators can fully enjoy the diverse and rewarding experiences that ham radio has to offer.

Different Classes of Licenses

Amateur radio licenses are divided into different classes, each offering varying levels of privileges and responsibilities. In the United States, there are three main classes of licenses: Technician, General, and Amateur Extra. Each class builds upon the knowledge and skills required for the previous one, allowing operators to progressively access more frequencies and modes of communication.

The **Technician** class license is the entry-level license and serves as the starting point for most new ham radio operators. It requires passing a relatively straightforward exam that covers basic electronics, operating practices, and regulations. With a Technician license, operators gain access to the VHF and UHF bands, which are ideal for local communications. This includes using repeaters, participating in local nets, and experimenting with digital modes such as D-STAR and APRS.

Advancing to the **General** class license involves a more comprehensive examination that includes additional topics in radio theory, operating practices, and regulations. The General class license opens up a wide range of HF (high frequency) bands, allowing operators to communicate over long distances. This expanded access enables operators to engage in international contacts, participate in contests, and explore various modes such as CW (Morse code) and digital modes like PSK31 and FT8.

The **Amateur Extra** class license is the highest level of amateur radio licensing. It requires passing a rigorous exam that covers advanced topics in electronics, propagation, and regulatory practices. With an Amateur Extra license, operators gain access to all amateur radio bands and modes, including exclusive segments of the HF bands. This top-tier license reflects a high level of expertise and commitment to the hobby, allowing operators to fully explore the diverse and technical aspects of amateur radio.

Each class of license offers unique opportunities and challenges, encouraging operators to continuously expand their knowledge and skills. By advancing through the license classes, operators can enjoy a broader range of experiences and contribute more effectively to the ham radio community.

The Licensing Process

Obtaining an amateur radio license involves a structured process that ensures operators have the necessary knowledge and skills. The process typically includes studying for the exam, finding a testing session, taking the exam, and receiving your license.

1. **Studying for the Exam**: The first step in obtaining a license is to prepare for the exam. There are many resources available to help you study, including books, online courses, and practice exams. The ARRL and other organizations offer study guides that cover the material in each license class. It's important to focus on understanding the concepts, as well as memorizing the specific rules and regulations.

2. **Finding a Testing Session**: Once you feel prepared, you'll need to find a testing session. The FCC-authorized Volunteer Examiner Coordinators (VECs) organize these sessions, which are conducted by Volunteer Examiners (VEs). You can find a testing session through the ARRL website or local ham radio clubs. Many testing sessions are held in person, but some VECs also offer remote testing options.

3. **Taking the Exam**: On the day of the exam, you'll need to bring identification and any required fees. The exam consists of multiple-choice questions that cover the material for the specific license class. The Technician exam has 35 questions, the General exam has 35 questions, and the Amateur Extra exam has 50 questions. To pass, you need to answer at least 74% of the questions correctly.

4. **Receiving Your License**: After passing the exam, your results are submitted to the FCC, which will issue your license. This process can take a few days to a few weeks. Once your

license is granted, you will be assigned a call sign and can begin operating on the frequencies and modes allowed by your license class.

The licensing process is designed to be accessible and educational, ensuring that new operators are well-prepared to join the ham radio community. By following these steps and dedicating time to study, you can obtain your license and start enjoying the many benefits of amateur radio.

Rules and Regulations

Amateur radio operates under a set of rules and regulations designed to ensure the efficient and safe use of the radio spectrum. These regulations are established by the FCC in the United States and are detailed in Part 97 of the FCC Rules. Understanding these rules is crucial for all amateur radio operators, as they govern everything from frequency allocations to operating practices.

One of the primary purposes of these regulations is to prevent interference with other communication services. Amateur radio operators must adhere to frequency allocations and power limits to avoid causing harmful interference. Specific bands are allocated for amateur use, and operators must use these bands according to the privileges granted by their license class.

Operators are also required to identify themselves regularly during transmissions using their call sign. This practice ensures accountability and allows other operators to know who is on the air. The call sign must be transmitted at the beginning and end of each communication, as well as at regular intervals during longer transmissions.

Another important aspect of the regulations is the prohibition of commercial use. Amateur radio is strictly a non-commercial service, and operators are not allowed to conduct business or transmit commercial advertisements. This helps maintain the hobby's focus on personal communication, experimentation, and public service.

Emergency communication is a significant part of amateur radio, and the regulations provide special provisions for it. In times of emergency, when normal communication channels are unavailable, amateur radio operators can provide critical links for emergency services and disaster relief efforts. These operations must be conducted in accordance with established emergency communication protocols and within the framework of the rules.

Operators must also respect the privacy of other communications. Interception and disclosure of non-amateur communications are prohibited. This includes refraining from listening to or sharing private conversations that occur on frequencies not allocated for amateur use.

In addition to national regulations, amateur radio operators must be aware of international rules and agreements. The ITU sets global standards for frequency allocations and operating practices, ensuring that amateur radio can function smoothly across borders. Operators who wish to communicate internationally must follow these guidelines to avoid causing interference and to promote international cooperation.

Staying informed about the latest rules and regulations is an ongoing responsibility for all amateur radio operators. The ARRL and other organizations provide regular updates and resources to help operators stay compliant. By adhering to these regulations, operators contribute to the orderly and efficient use of the radio spectrum, ensuring that the amateur radio service remains a valuable resource for communication and public service.

FCC Rules and Regulations

The FCC's rules and regulations for amateur radio are comprehensive and cover a wide range of topics. These rules are codified in Part 97 of the FCC Rules and include guidelines on licensing, operating procedures, and technical standards. Familiarity with these rules is essential for all amateur radio operators.

One of the key aspects of the FCC rules is the allocation of frequency bands. The FCC assigns specific portions of the radio spectrum for amateur use, and these bands are divided into different segments for various modes of operation, such as voice, digital, and Morse code. Operators must use the frequencies allocated to their license class and follow the mode restrictions for each band segment.

The FCC also establishes power limits for amateur radio transmissions. These limits vary depending on the frequency band and the mode of operation. Adhering to power limits helps prevent interference with other communication services and ensures that amateur radio signals do not disrupt other users of the radio spectrum.

Another critical component of the FCC rules is the requirement for station identification. Operators must transmit their call sign at the beginning and end of each communication and at regular intervals during extended transmissions. This practice promotes accountability and transparency within the amateur radio community.

The FCC rules also address the use of encryption and coded messages. In general, amateur radio communications must be open and accessible to all operators. The use of encryption is prohibited, except for certain specific purposes such as controlling space stations or satellites. This ensures that amateur radio remains a transparent and cooperative service.

In addition to these technical and operational guidelines, the FCC rules emphasize the non-commercial nature of amateur radio. Operators are prohibited from using their stations for commercial purposes, including transmitting advertisements or conducting business transactions. This restriction helps maintain the hobby's focus on personal communication, experimentation, and public service.

Emergency communication is another important area covered by the FCC rules. During emergencies, amateur radio operators are authorized to provide essential communication links, even if it means using frequencies and power levels outside their usual privileges. This flexibility is crucial for supporting disaster response and public safety efforts.

The FCC rules also include provisions for international communication. Operators who wish to communicate with stations in other countries must follow international agreements and guidelines established by the ITU. This helps ensure that amateur radio can function effectively on a global scale and fosters international goodwill.

Overall, the FCC rules and regulations are designed to promote the responsible and efficient use of the radio spectrum. By understanding and adhering to these rules, amateur radio operators can ensure that their activities are safe, compliant, and beneficial to the broader community.

Operating Privileges and Restrictions

Operating privileges in amateur radio vary depending on the class of license held by the operator. Each class of license grants access to specific frequency bands and modes of communication, with higher classes offering more extensive privileges.

Technician class licensees have access to all VHF and UHF bands, which are ideal for local and regional communication. These bands include popular frequencies such as the 2-meter and 70-centimeter bands. Technicians can use FM, digital modes, and even some limited HF privileges, including parts of the 10-meter band for voice and digital modes.

General class licensees enjoy significantly expanded privileges, including access to most HF bands. This allows for long-distance communication across continents. Generals can operate on all VHF and UHF bands as well, and they have privileges on the majority of the HF bands, enabling them to participate in international contacts, contests, and a variety of communication modes.

Amateur Extra class licensees have the most comprehensive privileges, with access to all amateur radio bands and modes. This includes exclusive segments of the HF bands reserved for Extra class operators, providing more opportunities for experimentation and communication. Extras can operate on all VHF, UHF, and HF frequencies, giving them the fullest range of operating possibilities.

Each class of license also comes with specific restrictions. For example, operators must adhere to power limits, which vary depending on the band and mode. These limits are set to prevent interference with other communication services and to ensure safe operation.

Additionally, operators must follow band plans, which designate specific frequencies for different modes of operation. This helps organize the use of the radio spectrum and prevents conflicts between different types of communications.

Understanding and respecting these privileges and restrictions is essential for responsible and effective operation in the amateur radio bands. By doing so, operators can fully enjoy the diverse and dynamic world of ham radio while contributing to the orderly and efficient use of the radio spectrum.

International Regulations

Amateur radio is governed by international regulations established by the ITU, a specialized agency of the United Nations. These regulations ensure that amateur radio operates harmoniously across borders, promoting global cooperation and minimizing interference.

The ITU allocates frequency bands for amateur use and sets guidelines for their operation. These allocations are outlined in the ITU Radio Regulations, which provide a framework for national authorities to implement and enforce. Operators who wish to communicate internationally must adhere to these guidelines, which include respecting frequency allocations and operating standards.

International agreements also facilitate cross-border communication by establishing reciprocal licensing arrangements. These agreements allow licensed operators to operate in other countries without obtaining a new license, fostering international collaboration and exchange.

By following international regulations, amateur radio operators contribute to a global community of enthusiasts who share a commitment to responsible and efficient use of the radio spectrum.

Call Signs

Call signs are unique identifiers assigned to licensed amateur radio operators. They serve as an official means of identification during communications, ensuring accountability and promoting orderly use of the radio spectrum.

In the United States, call signs are issued by the FCC and follow a specific format. A typical call sign consists of a prefix, a numeral, and a suffix. For example, the call sign "K1ABC" includes the prefix "K," the numeral "1," and the suffix "ABC." The prefix and numeral often indicate the geographic region of the operator, while the suffix is a unique identifier.

Call signs are issued sequentially based on the license class and the operator's location. Higher license classes, such as General and Amateur Extra, may be eligible for shorter call signs with fewer characters, which are often considered more prestigious.

Operators must use their call signs at the beginning and end of each communication, as well as at regular intervals during extended transmissions. This practice ensures that all communications are traceable and that operators can be identified easily.

Special event call signs and vanity call signs add another layer of personalization and recognition. Special event call signs are temporary and often used to commemorate significant events or achievements. Vanity call signs, on the other hand, are permanent and allow operators to choose a call sign that is meaningful or personalized, subject to availability and regulatory approval.

Understanding and using call signs correctly is a fundamental aspect of amateur radio operation. It enhances the professionalism and integrity of the hobby, ensuring that all operators are accountable and easily identifiable during their communications.

Structure and Allocation

The structure and allocation of amateur radio frequency bands are essential for organized and efficient communication. The FCC and the ITU allocate specific portions of the radio spectrum for amateur use, dividing them into different bands based on frequency.

Each band is designated for various modes of operation, such as voice, digital, and Morse code. This segmentation helps prevent interference between different types of communications and ensures that operators can find clear frequencies for their transmissions.

For example, the HF bands are allocated for long-distance communication, while the VHF and UHF bands are used for local and regional communication. Each band has specific sub-bands for different modes, ensuring that voice, digital, and CW operators can coexist without causing mutual interference.

Adhering to these allocations is crucial for maintaining the orderly use of the radio spectrum. By following band plans and respecting frequency allocations, operators can enjoy effective and interference-free communication across the diverse range of amateur radio bands.

Vanity Call Signs

Vanity call signs are a unique feature of amateur radio, allowing operators to choose a personalized call sign that reflects their identity or interests. This option is available to licensed operators who meet certain criteria and are willing to pay a fee.

The process of obtaining a vanity call sign involves submitting an application to the FCC, specifying the desired call sign. The chosen call sign must be available and comply with the format rules set by the FCC. Vanity call signs can be shorter and more distinctive than sequentially issued call signs, making them desirable for operators who want a unique and memorable identifier.

Vanity call signs can commemorate significant achievements, reflect personal interests, or simply provide a more distinctive identity on the airwaves. They add a personal touch to the hobby and allow operators to express their individuality within the amateur radio community.

By understanding the rules and procedures for obtaining vanity call signs, operators can enhance their ham radio experience and enjoy a personalized and distinctive presence on the air.

CHAPTER 2 – Operating Practices

Basic Operating Procedures

Mastering basic operating procedures is fundamental for every ham radio operator. These practices ensure efficient communication and promote good etiquette on the airwaves. Whether you're making your first contact or participating in a net, following these procedures will enhance your experience and help you become a respected operator.

One of the first things to learn is how to properly initiate a contact, known as a "QSO" (from the Q-code meaning "I can communicate with..."). Start by listening to the frequency to ensure it's not in use. This practice, called "listening before transmitting," helps prevent accidental interference with ongoing communications. When you're ready, you can announce your presence by saying "CQ" followed by your call sign. For example, "CQ CQ CQ, this is K1ABC calling CQ." Repeat this a few times, then listen for any response.

When someone responds to your CQ call, acknowledge their call sign and provide a brief signal report. Signal reports typically use the RST system, which stands for Readability, Strength, and Tone. Readability ranges from 1 (unreadable) to 5 (perfectly readable), strength from 1 (faint) to 9 (very strong), and tone (for CW) from 1 (very rough) to 9 (perfect tone). For voice modes, you'll often only use the first two numbers. An example exchange might be, "K1XYZ, this is K1ABC. You are 59 (five-nine) in New York. How copy?"

After exchanging signal reports, you can proceed with your conversation. Keep transmissions concise and avoid long-winded monologues. This allows for more interactive and engaging communication. During the QSO, it's important to listen carefully and acknowledge what the other operator says. This demonstrates good operating practice and helps build rapport.

Ending a QSO is as important as starting one. Conclude your conversation with a courteous sign-off, such as "73" (a common ham radio code meaning "best regards"). Be sure to repeat your call sign, so other operators know you are finished. For example, "Thank you for the contact, 73, this is K1ABC clear."

Another fundamental aspect of operating is participating in nets. Nets are organized on-air meetings where operators gather to discuss topics, pass traffic, or conduct drills. There are various types of nets, including traffic nets, emergency nets, and social nets. To join a net, listen to the net control station (NCS) and follow their instructions. When it's your turn, respond promptly and provide the requested information. Good net etiquette includes waiting for your turn to speak, being concise, and following the net's protocols.

Using repeaters is common in VHF and UHF operations. Repeaters extend the range of your transmission by receiving on one frequency and retransmitting on another. When using a repeater, key your microphone and wait a second before speaking to ensure your entire message is transmitted. Identify yourself with your call sign at the beginning and end of each transmission, and remember to leave a brief pause between exchanges to allow others to join the conversation.

Operating practices also include proper use of phonetics, especially in noisy conditions or when signal strength is weak. The standard phonetic alphabet (Alpha, Bravo, Charlie, etc.) helps ensure clarity and reduces misunderstandings. For example, instead of saying "K1ABC," you would say "Kilo One Alpha Bravo Charlie." This practice is particularly important during contests, DXing, and emergency communications.

Maintaining a logbook is a valuable habit, even though it is no longer a legal requirement in many countries. Logging your contacts helps you track your activity, verify contacts for awards, and maintain a record of your operating history. Essential details to log include the date, time, frequency, mode, call sign of the station worked, and any relevant notes about the QSO.

Respecting others on the airwaves is paramount. Avoid using offensive language, making unnecessary noise, or monopolizing a frequency. Courtesy, patience, and a willingness to help others are hallmarks of a good operator. If you hear a new operator struggling, offer guidance and encouragement. Remember, everyone was a beginner once, and fostering a supportive environment benefits the entire ham radio community.

Adhering to these basic operating procedures will help you become a proficient and respected ham radio operator. By practicing good etiquette, clear communication, and respect for others, you will enhance your enjoyment of the hobby and contribute positively to the ham radio community.

Communication Protocols

Effective communication protocols are essential for ensuring clarity and efficiency in ham radio operations. These protocols include standard phrases, procedures, and etiquette that help facilitate smooth and understandable exchanges between operators.

One fundamental protocol is the use of clear and concise language. Avoid slang or jargon that might confuse other operators, especially those who are non-native English speakers. Instead, use standard terminology that is widely understood within the ham radio community.

When identifying yourself, always use your call sign. This not only ensures legal compliance but also helps other operators recognize and respond to you. During extended conversations, it's good practice to periodically repeat your call sign to remind others who you are. For example, after a few exchanges, you might say, "This is K1ABC."

Break-in communication is another important protocol. In this method, you listen between transmissions for a break in the conversation, indicating that someone else wishes to join. This is particularly useful during nets or busy frequencies. To signal your intention to speak, you can say "Break" or your call sign during a pause. The net control or current operator should acknowledge you and give you an opportunity to speak.

The use of signal reports is a standard protocol in ham radio. Signal reports help operators understand how well their signals are being received and allow them to make adjustments if necessary. The RST system (Readability, Strength, and Tone) is commonly used for this purpose. Providing accurate signal reports fosters better communication and helps maintain the integrity of the exchange.

Another key protocol is the use of prowords (procedural words) and phonetics. Prowords are standard words or phrases used to convey specific meanings clearly and concisely. For example, "Roger" indicates that a message was received and understood, while "Over" signals the end of a transmission and invites a response. Phonetics, as mentioned earlier, help ensure that call signs and other critical information are communicated accurately, especially in poor conditions.

During emergency communications, protocols become even more critical. Operators must follow established procedures to ensure efficient and effective communication. This includes using designated emergency frequencies, adhering to net control instructions, and prioritizing urgent traffic. Standardized message formats and clear language help prevent misunderstandings and ensure that vital information is conveyed accurately.

Respecting the order of communication is another important protocol. During nets, the net control station (NCS) manages the flow of communication. Operators should wait for the NCS to acknowledge them before speaking. This orderly process ensures that everyone has a chance to participate and that messages are relayed efficiently.

In summary, adhering to communication protocols is essential for effective ham radio operation. These protocols include the use of clear language, proper identification, break-in communication, accurate signal reports, prowords, phonetics, and respecting the order of communication. By following these protocols, operators can enhance their communication skills, contribute to efficient use of the airwaves, and foster a positive and organized ham radio community.

Q-Signals and Common Abbreviations

Q-signals and common abbreviations are widely used in ham radio to facilitate quick and efficient communication. These shorthand codes help convey complex messages in a concise and standardized format, making it easier for operators to understand each other, especially in noisy or challenging conditions.

Q-signals are three-letter codes that start with the letter "Q." They are used to ask questions or provide information in a standardized way. Here are some commonly used Q-signals:

- **QTH**: Location. For example, "My QTH is New York" means "My location is New York."

- **QRZ**: Who is calling me? This signal is often used when an operator hears someone calling but needs to identify them.

- **QSL**: Can you acknowledge receipt? It is also used to confirm a contact, as in "QSL card," a postcard exchanged to confirm a QSO.

- **QRM**: Interference from other signals. "I am experiencing QRM" means there is interference from other signals.

- **QRN**: Interference from natural sources. "I am experiencing QRN" refers to noise from natural sources like lightning.

- **QRP**: Low power operation. "I am QRP" means the operator is using low power, typically 5 watts or less.

- **QRO**: High power operation. "I am QRO" means the operator is using higher power.

- **QSY**: Change frequency. "Let's QSY to 14.250 MHz" means switch to the frequency 14.250 MHz.

- **QRT**: Stop transmitting. "I am going QRT" means the operator is going off the air.

Common abbreviations also play a vital role in ham radio communications. These abbreviations are often derived from the first few letters of a word or a phonetic representation. Here are some frequently used abbreviations:

- **73**: Best regards. Used to sign off at the end of a QSO.

- **88**: Love and kisses. A friendly sign-off used among friends or family.

- **CQ**: Calling any station. Used to initiate a call to any operator listening.

- **DX**: Distance. Refers to long-distance communication or a distant station.
- **YL**: Young lady. Refers to a female operator.
- **OM**: Old man. Refers to a male operator.
- **SK**: Silent Key. Refers to a deceased ham radio operator.

Using Q-signals and abbreviations effectively requires familiarity and practice. They help streamline communication and reduce the time spent transmitting, which is especially important in high-traffic situations or when operating under challenging conditions.

In addition to these, CW (Morse code) operators often use prosigns, which are combinations of characters sent together as a single signal. Examples include:

- **AR**: End of message.
- **SK**: End of contact.
- **BT**: Separation between thoughts or sentences.
- **KN**: Invitation for a specific station to transmit.

Understanding and using Q-signals, abbreviations, and prosigns can greatly enhance your efficiency and effectiveness as a ham radio operator. These tools allow for clear, concise, and standardized communication, making it easier to convey information quickly and accurately. By incorporating these shorthand codes into your operating practices, you can improve your communication skills and enjoy a more streamlined and effective experience on the air.

Advanced Operating Techniques

As you gain experience in ham radio, you may want to explore advanced operating techniques that can enhance your communication skills and expand your capabilities. These techniques include contesting, DXing, portable operations, and special event stations.

Contesting is a popular activity where operators compete to make the most contacts in a given time period. Contests can range from a few hours to several days and may focus on specific bands, modes, or types of contacts. To be successful in contesting, operators need to develop skills in rapid and accurate communication, efficient logging, and effective station setup. Contesting not only hones your operating skills but also provides opportunities to connect with operators worldwide and achieve awards and recognition.

DXing, or long-distance communication, is another exciting aspect of ham radio. DXers strive to make contacts with stations in as many different countries and regions as possible. This pursuit often involves tracking propagation conditions, using directional antennas, and mastering various modes of communication. DXing can be particularly rewarding during periods of high solar activity when propagation conditions are favorable for long-distance contacts.

Portable operations involve setting up and operating your station away from your home base. This can include activities such as field days, Summits on the Air (SOTA), and Parks on the Air (POTA). Portable operations require careful planning and preparation, including selecting appropriate equipment, ensuring power sources, and considering environmental factors. These operations offer unique challenges and opportunities to explore different locations and test your skills in diverse conditions.

Special event stations are temporary stations set up to commemorate significant events, anniversaries, or achievements. These stations often use special call signs and may offer commemorative QSL cards or certificates. Operating a special event station can be a fun and rewarding way to engage with the ham radio community and promote interest in the hobby.

In addition to these activities, advanced operators often explore various modes of communication beyond traditional voice and CW. These modes include:

- **Digital modes** such as PSK31, FT8, and RTTY, which use computers and software to transmit and receive data. Digital modes are highly efficient and can achieve successful communication even in poor conditions.

- **Satellite communication**, which involves using amateur radio satellites to relay signals. This mode allows for contacts over long distances and can be an exciting way to explore the technical aspects of ham radio.

- **Moonbounce (EME)**, where operators bounce signals off the moon's surface to achieve long-distance communication. This highly technical mode requires precise equipment and skills but offers a unique and rewarding experience.

Developing proficiency in these advanced techniques requires continuous learning and practice. Many resources are available, including books, online forums, and clubs dedicated to specific aspects of ham radio. Participating in local and online communities can provide valuable insights, mentorship, and opportunities to collaborate with experienced operators.

Advanced operating techniques enhance your ham radio experience by introducing new challenges and expanding your horizons. Whether you're competing in contests, chasing DX, operating portable, or exploring digital modes, these techniques offer endless opportunities for growth and enjoyment in the hobby. Embracing these advanced practices will help you become a more versatile and skilled operator, capable of tackling a wide range of communication challenges and contributing to the vibrant and diverse ham radio community.

Digital Modes

Digital modes are a fascinating and rapidly growing aspect of ham radio that combine traditional radio communication with modern computer technology. These modes offer efficient and reliable ways to transmit data, making them ideal for various applications, from casual QSOs to emergency communications.

One of the most popular digital modes is **PSK31 (Phase Shift Keying, 31 baud)**. PSK31 is known for its ability to operate effectively in low signal-to-noise conditions, making it ideal for weak signal communications. It uses a narrow bandwidth of about 31 Hz, allowing multiple signals to coexist on the same frequency without significant interference. PSK31 is particularly popular for casual chats and digital contesting, and it is supported by many software programs, such as FLdigi and Ham Radio Deluxe.

FT8 (Franke-Taylor 8) is another widely used digital mode, developed by Joe Taylor (K1JT) and Steve Franke (K9AN). FT8 is designed for weak signal communication and can decode signals that are 20 dB below the noise level. It uses eight-tone frequency shift keying (FSK) and operates in 15-second transmission cycles, making it highly efficient for quick exchanges. FT8 is especially popular for DXing and band openings, as it allows operators to make contacts even in challenging propagation conditions. The WSJT-X software is commonly used for FT8 and other weak signal modes.

RTTY (Radio Teletype) is one of the oldest digital modes, dating back to the early 20th century. RTTY uses frequency shift keying (FSK) to transmit text data and is typically operated at 45.45 baud. Despite being an older mode, RTTY remains popular for contesting and digital communications. Many operators appreciate its straightforward, text-based nature and the excitement of participating in RTTY contests.

JT65 and JT9 are other digital modes developed by Joe Taylor, designed for weak signal communications like FT8. JT65 is known for its use in moonbounce (EME) and meteor scatter communications due to its robustness and ability to decode very weak signals. JT9 is similar but uses a narrower bandwidth, making it suitable for crowded bands. Both modes operate with one-minute transmission cycles, which require precise time synchronization, typically achieved using network time protocol (NTP) software.

OLIVIA is a digital mode designed for reliable text communication in poor conditions, including low signal-to-noise ratios and high levels of interference. OLIVIA uses multi-tone frequency shift keying (MFSK) and can be configured with different numbers of tones and bandwidths to suit various conditions. It is especially popular for ragchewing and emergency communications due to its robustness and ease of use.

Pactor is a digital mode often used for long-distance email and data transfer over HF bands. It combines aspects of packet radio and AMTOR (a precursor to RTTY) to provide reliable and error-free communication. Pactor is commonly used by maritime operators and remote stations for accessing email and internet services via HF.

Setting up for digital modes typically involves connecting your radio to a computer using an audio interface, such as a sound card or dedicated digital mode interface. Software programs like FLdigi, WSJT-X, and Ham Radio Deluxe provide the necessary tools to encode and decode digital signals, manage contacts, and log QSOs.

Digital modes offer a versatile and efficient way to communicate, opening up new possibilities for ham radio operators. Whether you're interested in weak signal work, casual conversations, or digital contesting, these modes provide exciting opportunities to explore and expand your skills. Embracing digital modes can enhance your ham radio experience, allowing you to communicate effectively even in challenging conditions and stay at the forefront of technological advancements in the hobby.

Satellite Communication

Satellite communication is an exciting and challenging aspect of ham radio that allows operators to make contacts via amateur radio satellites orbiting the Earth. These satellites, often referred to as "birds," provide a unique platform for long-distance communication and technical experimentation.

Amateur radio satellites are typically built and launched by organizations such as AMSAT (Radio Amateur Satellite Corporation) and various international ham radio groups. These satellites operate on designated amateur radio frequencies and can be used for voice, data, and digital communications. Some of the most popular satellites include AO-91, SO-50, and the International Space Station (ISS), which hosts a ham radio station.

To get started with satellite communication, you'll need some specialized equipment and knowledge. Here's a step-by-step guide:

1. **Equipment**: A dual-band VHF/UHF transceiver is essential for satellite work, as most satellites operate on these bands. You'll also need an antenna system capable of receiving and transmitting signals in both VHF and UHF. Directional antennas, such as Yagis or log periodic antennas, are commonly used to track and communicate with satellites. Additionally, a rotator system can help you aim your antennas precisely as the satellite moves across the sky.

2. **Tracking**: Satellite communication requires precise tracking of the satellite's position and movement. Tracking software, such as SatPC32, Orbitron, or the AMSAT website, provides real-time information on satellite orbits, pass times, and frequencies. These tools help you know when a satellite is within range and where to point your antennas.

3. **Doppler Shift**: As a satellite moves relative to your position, the frequency of its signal will shift due to the Doppler effect. You'll need to adjust your transceiver's frequency during the pass to compensate for this shift. Tracking software often includes Doppler correction features to help you make these adjustments accurately.

4. **Operating**: Begin by listening to the satellite's downlink frequency to ensure it's active and to identify any ongoing communications. When the satellite is within range, transmit on the uplink frequency and call "CQ satellite" followed by your call sign. Be concise and clear in your transmissions, as satellite passes are brief, typically lasting only 10 to 15 minutes.

5. **Etiquette**: Satellite communication requires good operating etiquette. Be mindful of other operators and avoid monopolizing the frequency. Wait for a clear moment to transmit and keep your exchanges brief to allow others a chance to make contacts.

6. **Experimentation**: Satellite work offers many opportunities for experimentation, including building your own antennas, developing tracking systems, and exploring digital modes. The technical challenges of satellite communication can be very rewarding and provide valuable learning experiences.

7. **International Space Station (ISS)**: The ISS hosts an amateur radio station that is occasionally available for contacts with Earth-bound operators. Astronauts aboard the ISS often participate in scheduled ham radio contacts, known as ARISS (Amateur Radio on the International Space Station) events. These contacts provide a unique opportunity to communicate with astronauts and engage in educational outreach.

Satellite communication adds an exciting dimension to ham radio, combining technical challenges with the thrill of making contacts across great distances. By learning the skills and techniques required for satellite operation, you can expand your capabilities and enjoy a unique aspect of the hobby. Whether you're aiming to make your first satellite contact or exploring advanced satellite experiments, this mode offers endless opportunities for growth and exploration in the world of ham radio.

Emergency Communications

Emergency communications (EmComm) is a vital aspect of ham radio, providing essential communication links during disasters and emergencies when conventional communication systems fail. Ham radio operators play a critical role in supporting emergency response efforts, ensuring that vital information is transmitted efficiently and reliably.

Preparation is key to effective emergency communications. Operators should have a well-equipped station with reliable power sources, such as batteries, solar panels, or generators. Portable and mobile setups are particularly valuable, allowing operators to establish communication from remote or affected areas. Ensuring your equipment is regularly tested and maintained is essential for readiness.

Training and Certification: Many organizations offer training and certification programs for emergency communications. The ARRL's Amateur Radio Emergency Service (ARES) provides training in emergency communication protocols, message handling, and net operations. Other organizations, such as the Radio Amateur Civil Emergency Service (RACES) and the Community Emergency Response Team (CERT), also offer valuable training programs. Obtaining certifications from these programs enhances your skills and ensures you are prepared to assist in an emergency.

Net Operations: During emergencies, communication is often organized through nets. A net is a structured on-air meeting led by a net control station (NCS) that coordinates the flow of information. Emergency nets can be local, regional, or national, depending on the scope of the emergency. Operators should familiarize themselves with net procedures, including how to check in, pass traffic, and follow the NCS's instructions. Efficient net operations ensure that messages are transmitted clearly and promptly.

Message Handling: Effective message handling is crucial in emergency communications. Messages should be concise, clear, and accurate. Standard message formats, such as the ARRL Radiogram, help ensure consistency and reduce the likelihood of errors. When passing messages, operators should use phonetics for critical information and confirm receipt to avoid misunderstandings.

Resource Management: In an emergency, managing available resources effectively is essential. This includes coordinating with other operators, sharing equipment, and utilizing available frequencies efficiently. Operators should be prepared to adapt to changing conditions and prioritize communication based on the needs of the emergency response.

Collaboration with Agencies: Ham radio operators often work closely with emergency response agencies, such as the Red Cross, FEMA, and local government agencies. Building relationships with these organizations before an emergency occurs can improve coordination and effectiveness. Understanding the needs and protocols of these agencies helps operators provide relevant and timely support.

Public Service Events: Participating in public service events, such as marathons, parades, and community fairs, is an excellent way to practice emergency communication skills. These events provide real-world experience in setting up stations, operating under varying conditions, and handling message traffic. They also offer opportunities to collaborate with local agencies and other ham radio operators.

EmComm Tools and Technology: Various tools and technologies enhance emergency communications. Digital modes, such as Winlink, provide reliable email and message capabilities over radio frequencies. APRS (Automatic Packet Reporting System) allows real-time tracking and messaging. Mesh networks, like AREDN (Amateur Radio Emergency Data Network), enable high-speed data communication in emergency situations. Familiarizing yourself with these tools and integrating them into your emergency communication plan can significantly enhance your capabilities.

Post-Emergency Evaluation: After an emergency, conducting a thorough evaluation of your communication efforts is important. Identify what worked well and areas for improvement. Debriefing with other operators and agencies involved in the response provides valuable insights and helps refine future emergency communication plans.

In summary, emergency communications is a critical component of ham radio, requiring preparation, training, and effective execution. By honing your skills, building relationships with agencies, and staying informed about the latest tools and technologies, you can be a valuable asset in times of crisis. Your contributions as a ham radio operator can make a significant difference in ensuring the safety and well-being of your community during emergencies.

Preparedness and Protocols

Being prepared for emergencies is a fundamental aspect of effective communication. As a ham radio operator, your readiness can make a crucial difference during a crisis. Preparedness involves having the right equipment, understanding emergency protocols, and maintaining a state of readiness to respond quickly and effectively.

Equipment Readiness: Ensure your radio equipment is in good working order and capable of operating under various conditions. This includes having backup power sources such as batteries, solar panels, and generators. Portable and mobile setups are essential for deployment in the field. Regularly test and maintain your equipment to ensure reliability.

Go-Kit: Assemble a go-kit containing all the essential items you'll need for emergency operations. This kit should include your radio, power supplies, antennas, coaxial cables, connectors, spare batteries, a first aid kit, personal items, and any necessary documentation. Having a go-kit ready allows you to quickly mobilize and set up your station in an emergency.

Training and Drills: Regular training and participation in emergency drills are vital for maintaining readiness. Join local ARES, RACES, or CERT groups to participate in training sessions and simulations. These exercises help you become familiar with emergency protocols, practice your skills, and coordinate with other operators and agencies.

Communication Plans: Develop and familiarize yourself with local and regional communication plans. These plans outline the frequencies, modes, and procedures to be used during emergencies. Having a clear communication plan ensures that you know where to tune and how to operate when normal communication channels are disrupted.

Protocols: Understand and follow established emergency communication protocols. This includes using designated emergency frequencies, adhering to net control instructions, and prioritizing traffic based on urgency. Efficient use of these protocols ensures that vital information is transmitted promptly and accurately.

Documentation: Keep important documentation handy, including frequency lists, net schedules, contact information for local emergency coordinators, and any relevant operating manuals. Having these resources readily available can save time and reduce confusion during an emergency.

Stay Informed: Monitor local news, weather reports, and emergency alerts to stay informed about potential emergencies. Being aware of developing situations allows you to anticipate and prepare for potential communication needs.

Personal Preparedness: Ensure your own safety and well-being so you can effectively assist others. This includes having an emergency plan for your family, securing your home, and maintaining adequate supplies of food, water, and medical supplies.

Post-Emergency Evaluation: After an emergency, take the time to evaluate your response and identify areas for improvement. Gather feedback from other operators and agencies involved in the response. Use this information to refine your preparedness plans and protocols.

Preparedness and adherence to protocols are essential for effective emergency communication. By maintaining your equipment, participating in training, and staying informed, you can be ready to respond when needed. Your readiness ensures that you can provide valuable communication support during emergencies, helping to protect and assist your community.

Participating in Emergency Nets

Participating in emergency nets is a crucial skill for ham radio operators, enabling effective coordination and communication during crises. Emergency nets are structured on-air meetings that facilitate the organized exchange of information and resources among operators and emergency response teams.

Net Control Station (NCS): The Net Control Station plays a pivotal role in managing the flow of communication. The NCS is responsible for maintaining order, directing traffic, and ensuring that all messages are transmitted clearly and efficiently. When participating in an emergency net, always follow the instructions of the NCS and wait for your turn to speak.

Checking In: To join an emergency net, listen carefully for a break in transmissions and then check in with your call sign. For example, you might say, "This is K1ABC checking in." The NCS will acknowledge your check-in and add you to the list of active participants. If you have traffic (messages) to pass, indicate this when you check in by saying, "This is K1ABC with traffic."

Passing Traffic: When it's your turn to pass traffic, be concise and clear. Use standard message formats and phonetics for critical information. For example, you might say, "This is K1ABC with a priority message for the Red Cross. The message reads: Shelter at 123 Main Street needs additional supplies. Over." Always confirm receipt of your message to ensure it has been understood correctly.

Priority Levels: Understand the different priority levels of messages. Emergency traffic (life-threatening situations) takes precedence, followed by priority traffic (important but not life-threatening), and routine traffic (general information). The NCS will prioritize messages based on their urgency, so be sure to indicate the priority level of your traffic.

Listening: Active listening is crucial in emergency nets. Pay close attention to the NCS and other operators to avoid missing important information. Keep transmissions brief and to the point, and avoid unnecessary chatter. This ensures that the net remains focused and efficient.

Resource Coordination: Emergency nets often involve coordinating resources and support. Be prepared to provide information about available resources, such as shelter locations, medical supplies, or transportation options. Accurate and timely information can significantly impact the effectiveness of emergency response efforts.

Relaying Messages: In some cases, you may need to relay messages between stations that cannot communicate directly. Ensure you accurately copy the message and transmit it without alteration. Relaying messages helps extend the reach of the net and ensures that vital information is disseminated widely.

Net Etiquette: Good net etiquette is essential for maintaining order and efficiency. Always identify yourself with your call sign before transmitting, follow the NCS's instructions, and avoid interrupting other operators. If you need to leave the net, inform the NCS and check out with your call sign.

After-Action Reports: After the emergency net concludes, participate in any after-action reports or debriefings. These discussions provide valuable feedback and help identify areas for improvement. Sharing your experiences and insights contributes to the continuous improvement of emergency communication practices.

Participating in emergency nets requires discipline, clear communication, and adherence to established protocols. By mastering these skills, you can effectively contribute to emergency response efforts and support your community during times of crisis. Your participation in emergency nets ensures that vital information is exchanged efficiently, helping to coordinate resources and protect lives and property.

CHAPTER 3 – Radio Technology Fundamentals

Basic Electronics and Components

To understand radio technology, it's essential to start with the basics of electronics and components. These fundamentals form the backbone of all radio systems, from simple handheld transceivers to complex communication networks.

Electricity Basics: Electricity is the flow of electrons through a conductor, typically measured in amperes (amps). Voltage, measured in volts, is the potential difference that drives this flow. Resistance, measured in ohms, opposes the flow of electrons. These three elements form the foundation of electrical circuits.

Direct Current (DC) and Alternating Current (AC): In DC circuits, electrons flow in a single direction, while in AC circuits, the direction of flow alternates periodically. Radio equipment typically uses DC for internal circuitry, while AC is used for power distribution in households and buildings.

Electronic Components: Various components manipulate the flow of electricity in circuits. Here are some fundamental components:

- **Resistors**: Resistors oppose the flow of current and are used to control voltage and current levels within circuits. They are characterized by their resistance value, measured in ohms.

- **Capacitors**: Capacitors store and release electrical energy, smoothing voltage fluctuations and filtering signals. Their capacitance is measured in farads.

- **Inductors**: Inductors store energy in a magnetic field when current flows through them. They are commonly used in filtering and tuning circuits. Inductance is measured in henrys.

- **Diodes**: Diodes allow current to flow in only one direction, acting as a one-way valve. They are used for rectification, converting AC to DC.

- **Transistors**: Transistors amplify or switch electronic signals and are fundamental in modern electronics. They come in two main types: bipolar junction transistors (BJTs) and field-effect transistors (FETs).

Integrated Circuits (ICs): ICs are complex devices that contain multiple electronic components on a single silicon chip. They perform various functions, from amplification to digital processing.

Power Supplies: Power supplies convert and regulate electrical power for electronic devices. They can be linear or switching types, with switching power supplies being more efficient.

Schematic Diagrams: Schematic diagrams represent electronic circuits using symbols for components and lines for connections. Understanding these diagrams is crucial for designing and troubleshooting circuits.

Ohm's Law and Basic Circuits

Ohm's Law is a fundamental principle in electronics, describing the relationship between voltage (V), current (I), and resistance (R). It is expressed as:

$V = I \times R$

This law is vital for analyzing and designing electronic circuits. Here are some basic applications:

Series Circuits: In a series circuit, components are connected end-to-end, with the same current flowing through each component. The total resistance is the sum of individual resistances:

$R_{total} = R_1 + R_2 + R_3 + ...$

The total voltage across the circuit is the sum of the voltages across each component.

Parallel Circuits: In a parallel circuit, components are connected across the same voltage source, with different branches carrying different currents. The total resistance is given by:

$1/R_{total} = 1/R_1 + 1/R_2 + 1/R_3 + ...$

The total current is the sum of the currents through each branch.

Combination Circuits: These circuits combine series and parallel elements. Analyzing these circuits involves breaking them down into simpler series and parallel sections.

To illustrate, consider the following table of resistor combinations:

Configuration	Formula	Example Calculation
Series	$R_{total} = R_1 + R_2$	$R_{total} = 10\Omega + 20\Omega = 30\Omega$
Parallel	$\frac{1}{R_{total}} = \frac{1}{R_1} + \frac{1}{R_2}$	$\frac{1}{R_{total}} = \frac{1}{10\Omega} + \frac{1}{20\Omega}$
Combination	Solve series and parallel separately	-

Here's an example of a combination circuit:

1. **Identify series and parallel sections**: Break down the circuit into simpler parts.
2. **Calculate equivalent resistances**: Use the formulas for series and parallel resistances.
3. **Combine results**: Add up the equivalent resistances to find the total resistance.

Visual aids, such as circuit diagrams and simulation tools, can greatly enhance understanding.

The image above represents a simple series circuit, illustrating the concept of Ohm's Law in practice.

Components: Resistors, Capacitors, Inductors

Understanding the characteristics and applications of resistors, capacitors, and inductors is crucial for designing and troubleshooting radio circuits.

Resistors

- **Fixed Resistors**: These have a constant resistance value. They are used to limit current, divide voltages, and bias active elements.

- **Variable Resistors**: Also known as potentiometers or rheostats, these allow adjustable resistance. They are used for tuning and calibration.

- **Power Ratings**: Resistors have power ratings that indicate the maximum power they can dissipate without damage. It's essential to choose resistors with appropriate power ratings for your application.

Capacitors

- **Electrolytic Capacitors**: These have high capacitance values and are polarized, meaning they must be connected with the correct polarity. They are used for filtering and power supply smoothing.

- **Ceramic Capacitors**: These are non-polarized and suitable for high-frequency applications. They are commonly used in RF circuits for bypassing and coupling.

- **Tantalum Capacitors**: These offer high capacitance in a small size but are sensitive to over-voltage.

- **Applications**: Capacitors are used in timing circuits, filtering, signal coupling, and decoupling. They can store energy and release it when needed, which is crucial in power supply circuits.

Inductors

- **Fixed Inductors**: These have a constant inductance value. They are used in filters, chokes, and transformers.

- **Variable Inductors**: These allow the inductance to be adjusted, useful in tuning circuits.

- **Core Materials**: Inductors can have air, iron, or ferrite cores, which affect their inductance and frequency characteristics.

- **Applications**: Inductors store energy in a magnetic field and are used in filtering, tuning, and energy storage applications. They are essential in RF circuits for impedance matching and frequency selection.

To better understand these components, consider the following table:

Component	Symbol	Function	Example Application
Resistor	R	Limits current, divides voltage	Voltage divider, biasing transistor
Capacitor	C	Stores energy, filters signals	Power supply filtering, coupling
Inductor	L	Stores energy in magnetic field, filters	RF tuning, chokes

Understanding Radio Waves

Radio waves are a form of electromagnetic radiation used for wireless communication. Understanding their properties and behavior is fundamental to radio technology.

Electromagnetic Spectrum: Radio waves occupy a portion of the electromagnetic spectrum, ranging from about 3 kHz to 300 GHz. This spectrum is divided into different bands, such as LF, MF, HF, VHF, UHF, and microwave, each with unique propagation characteristics and applications.

Wave Properties:

- **Frequency (f)**: The number of cycles a wave completes in one second, measured in hertz (Hz).

- **Wavelength (λ)**: The distance a wave travels during one cycle, inversely related to frequency.

- **Amplitude**: The height of the wave, representing the signal strength.

- **Phase**: The position of the wave relative to a reference point, important in modulation and signal processing.

Propagation: Radio waves can propagate through space, air, and various materials. The mode of propagation depends on the frequency and environmental conditions.

- **Ground Wave**: Travels along the Earth's surface, useful for LF and MF bands.

- **Sky Wave**: Reflects off the ionosphere, enabling long-distance communication in HF bands.

- **Line-of-Sight**: Travels straight from transmitter to receiver, typical in VHF, UHF, and microwave bands.

Modulation: Modulation is the process of varying a carrier wave to transmit information. Common types include:

- **Amplitude Modulation (AM)**: Varies the amplitude of the carrier wave.

- **Frequency Modulation (FM)**: Varies the frequency of the carrier wave.

- **Phase Modulation (PM)**: Varies the phase of the carrier wave.

Understanding these concepts is crucial for designing and operating radio systems effectively.

Properties of Radio Waves

Radio waves exhibit several key properties that affect their behavior and suitability for different applications.

Reflection: Radio waves can bounce off surfaces, such as buildings, mountains, and the ionosphere. This property enables long-distance communication via sky wave propagation.

Refraction: Changes in the medium, such as atmospheric layers, can bend radio waves. Refraction is significant in VHF and higher frequencies, affecting signal paths.

Diffraction: Radio waves can bend around obstacles, allowing signals to reach areas not in the direct line of sight. This property is more pronounced at lower frequencies.

Absorption: Various materials can absorb radio waves, attenuating the signal. Moisture, vegetation, and buildings can absorb RF energy, impacting signal strength.

Polarization: The orientation of the electric field vector of a radio wave. Polarization can be linear (horizontal or vertical) or circular. Matching the polarization of the transmitter and receiver antennas is essential for optimal signal reception.

Interference: Unwanted signals from other sources can interfere with desired signals, causing noise and signal degradation. Interference can come from other transmitters, electronic devices, or natural sources.

Multipath: Multiple reflections and refractions can cause signals to take different paths to the receiver, leading to phase differences and signal fading. Techniques like diversity reception and spread spectrum help mitigate multipath effects.

Attenuation: The reduction of signal strength as it travels through space or materials. Attenuation increases with distance and frequency, requiring higher power or sensitive receivers to maintain communication.

Understanding these properties helps in designing efficient antennas, choosing appropriate frequencies, and mitigating potential issues in radio communication systems.

Frequency and Wavelength

Frequency and wavelength are fundamental concepts in understanding radio waves and their behavior.

Frequency (f): Frequency is the number of cycles a wave completes in one second, measured in hertz (Hz). It determines the wave's position in the electromagnetic spectrum and affects its propagation characteristics. Higher frequencies have shorter wavelengths and are used for applications requiring high bandwidth, such as VHF, UHF, and microwave communication.

Wavelength (λ): Wavelength is the distance a wave travels during one cycle, measured in meters. It is inversely related to frequency:

$\lambda = c/f$

where c is the speed of light 3×10^8 (approximately meters per second).

Band Designations: The radio spectrum is divided into different bands based on frequency. Each band has specific characteristics and uses:

- **LF (Low Frequency)**: 30 kHz to 300 kHz, used for navigation and time signals.
- **MF (Medium Frequency)**: 300 kHz to 3 MHz, used for AM broadcasting and maritime communication.
- **HF (High Frequency)**: 3 MHz to 30 MHz, used for shortwave broadcasting, amateur radio, and international communication.
- **VHF (Very High Frequency)**: 30 MHz to 300 MHz, used for FM broadcasting, television, and aviation communication.
- **UHF (Ultra High Frequency)**: 300 MHz to 3 GHz, used for television, mobile phones, and GPS.
- **Microwave**: Above 3 GHz, used for satellite communication, radar, and wireless networking.

Propagation Characteristics: Frequency and wavelength influence how radio waves propagate. Lower frequencies can travel longer distances and penetrate obstacles better but require larger antennas. Higher frequencies offer higher data rates and are better for line-of-sight communication but are more susceptible to attenuation and interference.

Understanding the relationship between frequency and wavelength is crucial for selecting the right frequency band for your application and designing effective antennas and transmission systems.

Propagation Basics

Propagation refers to the way radio waves travel from the transmitter to the receiver. Different propagation modes are influenced by frequency, environment, and atmospheric conditions.

Ground Wave Propagation: Used primarily in the LF and MF bands, ground waves travel along the Earth's surface and can cover significant distances. They are affected by the conductivity of the ground and are suitable for maritime and AM broadcast communications.

Sky Wave Propagation: In the HF band, radio waves can be reflected by the ionosphere, allowing for long-distance communication. The ionosphere consists of several layers that ionize differently depending on solar activity and time of day. Sky wave propagation is essential for international broadcasting, amateur radio, and military communication.

Line-of-Sight Propagation: VHF, UHF, and microwave frequencies rely on line-of-sight propagation, where the transmitter and receiver need a clear, unobstructed path. This mode is used for FM radio, television broadcasting, mobile communications, and satellite links. The curvature of the Earth limits line-of-sight communication to about 30-40 miles unless elevated antennas or repeaters are used.

Tropospheric Propagation: At VHF and above, radio waves can be refracted or scattered by the troposphere, extending their range beyond the horizon. This effect is more pronounced during temperature inversions and weather fronts.

Sporadic-E Propagation: Occurring mostly in the VHF band, sporadic-E propagation involves reflections from ionized patches in the E layer of the ionosphere. It allows for occasional long-distance VHF communication, particularly during the summer months.

Meteor Scatter: Radio waves can be reflected off the ionized trails left by meteors entering the Earth's atmosphere. This mode is used for brief, high-speed data transmissions in the VHF band.

Auroral Propagation: High-energy particles from solar flares can ionize the Earth's polar regions, creating auroras. Radio waves can reflect off these ionized regions, enabling VHF and UHF communication over long distances.

Satellite Communication: Satellites act as repeaters in space, relaying signals between distant points on Earth. This mode uses UHF and microwave frequencies and provides reliable communication for global positioning, weather monitoring, and long-distance communication.

Understanding propagation basics helps operators choose the right frequency and mode for their communication needs, optimize antenna placement, and predict signal behavior under different conditions.

CHAPTER 4 – Circuitry and Design

Circuit Analysis

Circuit analysis is a critical skill for anyone involved in ham radio, enabling you to understand, design, and troubleshoot electrical circuits effectively. This process involves examining the behavior of electrical circuits by studying the relationships between the different components and the flow of electric current.

Key Concepts in Circuit Analysis

1. **Voltage, Current, and Resistance**

- **Voltage (V)**: The electric potential difference between two points, driving the current through the circuit.

- **Current (I)**: The flow of electric charge, measured in amperes (A).

- **Resistance (R)**: The opposition to current flow, measured in ohms (Ω).

2. **Ohm's Law** Ohm's Law is fundamental to circuit analysis, expressed as $V = I \times R$. This relationship allows you to calculate one of the three quantities if the other two are known.

3. **Kirchhoff's Laws** Kirchhoff's Laws are essential for analyzing complex circuits:

    - **Kirchhoff's Current Law (KCL)**: The total current entering a junction equals the total current leaving the junction.

    - **Kirchhoff's Voltage Law (KVL)**: The sum of all voltages around a closed loop equals zero.

4. **Series and Parallel Circuits**

    - **Series Circuits**: Components connected end-to-end, with the same current flowing through each. The total resistance is the sum of individual resistances: $R_{total} = R_1 + R_2 + ... + R_n$.

    - **Parallel Circuits**: Components connected across the same two points, sharing the same voltage. The total resistance is given by the reciprocal formula: $1/R_{total} = 1/R_1 + 1/R_2 + ... + 1/R_n$.

5. **Thevenin's and Norton's Theorems** These theorems simplify the analysis of complex circuits:

    - **Thevenin's Theorem**: Any linear circuit can be replaced by an equivalent circuit consisting of a single voltage source and a series resistance.

    - **Norton's Theorem**: Similar to Thevenin's, but the equivalent circuit consists of a single current source and a parallel resistance.

6. **Superposition Theorem** This theorem states that the response in any branch of a linear circuit with multiple independent sources is the sum of the responses caused by each source acting alone.

Practical Applications of Circuit Analysis

1. **Designing Efficient Circuits** Mastering circuit analysis allows you to design circuits that meet specific requirements, such as minimizing power consumption or maximizing signal integrity.

2. **Troubleshooting** Circuit analysis techniques help identify issues in a malfunctioning circuit. By measuring voltages and currents and comparing them to expected values, you can locate faulty components or connections.

3. **Optimizing Performance** Fine-tuning a circuit's performance often involves adjusting component values or configurations based on analysis results to achieve desired outcomes like improved stability or enhanced frequency response.

Tools and Techniques

1. **Simulation Software** Tools like SPICE (Simulation Program with Integrated Circuit Emphasis) enable you to simulate and analyze circuits before physically building them, saving time and resources.

2. **Multimeters and Oscilloscopes**

    - **Multimeters**: Measure voltage, current, and resistance, providing immediate feedback on the circuit's state.

    - **Oscilloscopes**: Visualize electrical signals' waveforms, helping to understand AC circuits' behavior and detect unwanted noise or distortions.

3. **Network Analyzers** Useful for analyzing complex networks, especially in RF circuits, to understand impedance and frequency response.

4. **Prototyping Tools** Breadboards, perf boards, and soldering tools are essential for constructing and testing circuits, allowing for iterative design and troubleshooting.

By mastering circuit analysis, you'll have the foundation to explore more advanced topics in electronics and ham radio, such as AC and DC circuits, impedance, and practical circuit design, which will be covered in the subsequent sections of this chapter. Understanding these principles will enable you to build, analyze, and optimize circuits effectively, enhancing both your theoretical knowledge and practical skills.

AC and DC Circuits

In the world of ham radio and electronics, understanding the differences and applications of AC (Alternating Current) and DC (Direct Current) circuits is fundamental. These two types of current flow differently and are used in various components and systems, each with its unique characteristics and applications.

Direct Current (DC) Circuits

1. **Nature of DC**

    - **Definition**: DC is the unidirectional flow of electric charge. Electrons move steadily in one direction through the circuit.

- **Sources**: Common sources of DC include batteries, DC power supplies, and solar cells.
- **Applications**: DC is typically used in low-voltage applications like powering small electronic devices, LED lighting, and charging batteries.

2. **Characteristics of DC Circuits**
   - **Voltage and Current**: In a DC circuit, the voltage is constant, and the current flows in one direction.
   - **Circuit Components**: Typical components include resistors, capacitors (for smoothing purposes), and inductors.

3. **Analysis of DC Circuits**
   - **Ohm's Law**: $V = I \times R$ is used to calculate voltage, current, or resistance.
   - **Series and Parallel Configurations**: Similar to general circuit analysis, DC circuits can be arranged in series and parallel, affecting the total resistance and voltage distribution.

Alternating Current (AC) Circuits

1. **Nature of AC**
   - **Definition**: AC is the flow of electric charge that periodically reverses direction. The voltage in AC circuits varies sinusoidally with time.
   - **Sources**: Common sources of AC include power outlets (mains electricity), AC generators, and transformers.
   - **Applications**: AC is widely used for transmitting power over long distances, powering household appliances, and in radio frequency applications.

2. **Characteristics of AC Circuits**
   - **Voltage and Current**: In an AC circuit, both voltage and current alternate in direction, typically in a sinusoidal pattern.
   - **Frequency**: AC circuits operate at a specific frequency, measured in Hertz (Hz). In most countries, the mains frequency is either 50 Hz or 60 Hz.

3. **Analysis of AC Circuits**
   - **Impedance**: Unlike DC circuits, AC circuits have impedance, which includes resistance (R), inductive reactance (X_L), and capacitive reactance (X_C).
     - **Inductive Reactance (X_L)**: $X_L = 2\pi f L$, where f is the frequency and L is the inductance.

- **Capacitive Reactance (XC)**: $XC = 1/2\pi fC$, where C is the capacitance.
- **Impedance Calculation**: The total impedance (Z) in an AC circuit is a combination of resistance and reactance: $Z = \sqrt{R^2 + (X_L - X_C)^2}$.
- **Ohm's Law for AC**: $V = I \times Z$, where Z is the impedance.

4. **Power in AC Circuits**
   - **Real Power (P)**: The actual power consumed by the circuit, calculated as $P = VI\cos\theta$, where θ is the phase angle between voltage and current.
   - **Reactive Power (Q)**: Power stored and released by inductors and capacitors, calculated as $Q = VI\sin\theta$.
   - **Apparent Power (S)**: The combination of real and reactive power, calculated as $S = VI$.

Practical Considerations in AC and DC Circuits

1. **Safety**
   - **DC Circuits**: Generally safer at low voltages but can be hazardous at high currents or voltages.
   - **AC Circuits**: Mains electricity is dangerous and requires proper insulation, grounding, and circuit protection.

2. **Component Behavior**
   - **Resistors**: Behave similarly in both AC and DC circuits.
   - **Capacitors and Inductors**: React differently in AC circuits, introducing reactance that affects the circuit's overall impedance.

3. **Power Supplies**
   - **DC Power Supplies**: Include batteries, DC power adapters, and regulated DC power supplies.
   - **AC Power Supplies**: Include mains electricity, transformers, and AC generators.

4. **Conversion Between AC and DC**
   - **Rectification**: The process of converting AC to DC using diodes and rectifier circuits.
   - **Inversion**: The process of converting DC to AC using inverters, which is common in solar power systems and uninterruptible power supplies (UPS).

By understanding the differences and applications of AC and DC circuits, you'll be better equipped to design and work with a wide range of electronic and radio frequency systems. This knowledge forms the foundation for more advanced topics, such as impedance and reactance, which will be covered in the next section of this chapter.

Impedance and Reactance

Understanding impedance and reactance is crucial for anyone involved in ham radio and electronics, as these concepts significantly impact the performance and design of AC circuits. Impedance and reactance affect how AC signals behave in a circuit, influencing aspects such as signal strength, phase, and frequency response.

Impedance (Z)

1. **Definition**

    - Impedance is the total opposition that a circuit offers to the flow of alternating current (AC). It is a complex quantity that includes both resistance (R) and reactance (X).

    - Mathematically, impedance is represented as $Z=R+jX$, where j is the imaginary unit.

2. **Components of Impedance**

    - **Resistance (R)**: The real part of impedance, representing the opposition to current flow that converts electrical energy into heat.

    - **Reactance (X)**: The imaginary part of impedance, representing the opposition to current flow caused by inductors and capacitors, which store and release energy.

Reactance (X)

1. **Inductive Reactance (XL)**

    - **Definition**: Inductive reactance is the opposition to the change of current by an inductor in an AC circuit.

    - **Formula**: $XL=2\pi fL$, where f is the frequency of the AC signal and L is the inductance in henrys (H).

    - **Frequency Dependence**: Inductive reactance increases with frequency. Higher frequencies result in higher opposition to current change.

2. **Capacitive Reactance (XC)**

    - **Definition**: Capacitive reactance is the opposition to the change of voltage by a capacitor in an AC circuit.

    - **Formula**: $XC=1/2\pi fC$, where C is the capacitance in farads (F).

    - **Frequency Dependence**: Capacitive reactance decreases with frequency. Higher frequencies result in lower opposition to voltage change.

Calculating Impedance in AC Circuits

1. **Series Circuits**

    - **Total Impedance**: In a series circuit, the total impedance is the sum of the individual impedances: $Z_{total}=Z_1+Z_2+...+Z_n$.

    - **Example**: For a series circuit with a resistor (R) and an inductor (L), the total impedance is $Z=R+jX_L$.

2. **Parallel Circuits**

    - **Total Impedance**: In a parallel circuit, the total impedance is found using the reciprocal formula: $1/Z_{total}=1/Z_1+1/Z_2+...+1/Z_n$.

    - **Example**: For a parallel circuit with a resistor and a capacitor (C), the total impedance is $\frac{1}{Z} = \frac{1}{R} + j\frac{1}{X_C}$.

Phase Relationships

1. **Phase Angle (θ)**

    - **Definition**: The phase angle is the angle of the impedance vector, indicating the phase difference between the voltage and current in an AC circuit.

    - **Calculation**: $\theta=\arctan(X/R)$.

    - **Interpretation**: A positive phase angle indicates that the current lags the voltage (inductive), while a negative phase angle indicates that the current leads the voltage (capacitive).

2. **Phasor Diagrams**

    - **Definition**: Phasor diagrams graphically represent the relationships between voltage and current phasors in AC circuits.

    - **Usage**: These diagrams help visualize how impedance and reactance affect phase relationships, aiding in the design and analysis of AC circuits.

Practical Implications

1. **Tuning Circuits**

    - **Resonance**: Occurs when inductive reactance equals capacitive reactance ($X_L=X_C$), resulting in purely resistive impedance. Resonant circuits are used in filters and oscillators.

    - **Applications**: Tuning circuits to specific frequencies is essential in radio communications to select desired signals and reject unwanted ones.

2. **Impedance Matching**

    - **Purpose**: Ensures maximum power transfer between different stages of a circuit or between the circuit and an external load.
    - **Techniques**: Use of transformers, matching networks, or specific component values to match the source and load impedances.

3. **Filters and Networks**

    - **Low-Pass Filters**: Allow low-frequency signals to pass while attenuating high-frequency signals. Use inductors and capacitors to create desired frequency responses.
    - **High-Pass Filters**: Allow high-frequency signals to pass while attenuating low-frequency signals. Design considerations include component selection and configuration.

Measuring Impedance and Reactance

1. **Impedance Analyzers**

    - **Function**: Measure the impedance of a circuit or component across a range of frequencies.
    - **Applications**: Essential for characterizing components and circuits, especially in RF and high-frequency applications.

2. **LCR Meters**

    - **Function**: Measure inductance (L), capacitance (C), and resistance (R) directly.
    - **Usage**: Useful for testing and validating components before incorporating them into circuits.

Building and Troubleshooting Circuits

Building and troubleshooting circuits are essential skills for any ham radio operator. The ability to construct, test, and fix circuits ensures that your equipment operates reliably and efficiently. This section will cover the practical aspects of these activities, providing tips and techniques for successful circuit construction and troubleshooting.

Building Circuits

1. **Planning and Design**

    - **Schematic Diagrams**: Start with a clear and accurate schematic diagram. This blueprint guides you through the construction process, showing the connections and components.

- **Component Selection**: Choose components based on the circuit's requirements, considering factors such as voltage, current ratings, and tolerance levels.
- **Breadboarding**: Before final assembly, prototype your circuit on a breadboard. This allows for easy modifications and troubleshooting without soldering.

2. **Tools and Equipment**

    - **Soldering Iron**: Essential for making permanent electrical connections. Use a good quality soldering iron with appropriate tips for different tasks.
    - **Multimeter**: Used for measuring voltage, current, and resistance. An indispensable tool for both building and troubleshooting circuits.
    - **Oscilloscope**: Helps visualize electrical signals in the circuit, providing insights into the circuit's behavior, particularly in AC circuits.
    - **Wire Cutters and Strippers**: Necessary for preparing wires for connections.
    - **Tweezers and Pliers**: Useful for handling small components and making precise adjustments.

3. **Construction Techniques**

    - **Soldering**: Ensure good soldering practices to avoid cold joints and ensure reliable connections. Heat the joint adequately, apply solder to the joint, not the iron, and allow it to cool naturally.
    - **Component Placement**: Follow the schematic and place components in a logical and organized manner, minimizing the length of connecting wires to reduce potential issues like interference and resistance.
    - **Testing**: Continuously test your circuit during the construction process. Check each section before moving on to the next to ensure that everything is functioning correctly.

4. **Safety Considerations**

    - **Work in a Well-Ventilated Area**: Soldering produces fumes that can be harmful if inhaled.
    - **Use Safety Equipment**: Wear safety glasses to protect your eyes from solder splashes and use a fume extractor to remove harmful fumes.
    - **Double-Check Connections**: Ensure all connections are correct to prevent shorts and other issues that could damage components or cause injury.

Troubleshooting Circuits

1. **Systematic Approach**

- **Visual Inspection**: Start with a thorough visual inspection. Look for obvious issues like broken wires, poor solder joints, or damaged components.
- **Check Power Supply**: Ensure the power supply is functioning correctly and providing the correct voltage and current levels.
- **Signal Tracing**: Use a signal generator and oscilloscope to trace the signal through the circuit. Identify where the signal deviates from expected behavior to locate faults.

2. **Common Issues and Solutions**
    - **Cold Solder Joints**: Appear dull and may cause intermittent connections. Reheat and apply fresh solder to fix.
    - **Short Circuits**: Occur when two conductors accidentally touch, creating an unintended path for current. Identify and separate the conductors.
    - **Open Circuits**: Happen when there is a break in the circuit. Check for broken wires or component leads and repair as necessary.
    - **Component Failures**: Test individual components (resistors, capacitors, transistors) using a multimeter or LCR meter to ensure they are functioning correctly. Replace faulty components.

3. **Using Diagnostic Tools**
    - **Multimeter**: Measure voltage, current, and resistance to verify circuit operation and identify faults.
    - **Oscilloscope**: Visualize the signal at various points in the circuit to detect anomalies.
    - **Signal Generator**: Inject known signals into the circuit and observe the output to diagnose issues.

4. **Documenting and Learning**
    - **Keep a Log**: Document each step of the troubleshooting process, noting what you checked and the results. This helps in systematic troubleshooting and avoids repeating steps.
    - **Learn from Mistakes**: Each troubleshooting session is a learning opportunity. Analyze what went wrong and how you fixed it to improve your skills.

Practical Tips

1. **Component Handling**
    - **Static Sensitivity**: Handle sensitive components like ICs and transistors with care to avoid static damage. Use anti-static wristbands and mats.

- **Orientation**: Pay attention to the orientation of polarized components like electrolytic capacitors, diodes, and transistors.

2. **Circuit Testing**

    - **Incremental Testing**: Test the circuit in stages. Power up and test each section before moving on to the next.

    - **Isolation**: If a circuit is complex, isolate sections to troubleshoot specific areas without interference from other parts of the circuit.

3. **Maintenance**

    - **Regular Checks**: Perform regular checks on your circuits and equipment to ensure everything is in working order.

    - **Cleaning**: Keep your work area and circuit boards clean. Dust and debris can cause shorts and other issues.

Schematic Reading

Schematic reading is a fundamental skill for anyone involved in electronics and ham radio. Schematics are the blueprints of electronic circuits, providing a visual representation of the components and their connections. Understanding how to read and interpret schematics is crucial for designing, building, and troubleshooting circuits.

Components and Symbols

1. **Basic Components**

    - **Resistors**: Represented by a zigzag line or a rectangle. The value is typically indicated next to the symbol.

    - **Capacitors**: Two types are common in schematics: electrolytic capacitors (a short line and a long line) and non-polarized capacitors (two parallel lines).

    - **Inductors**: Represented by a series of loops or a rectangle with a line through it, indicating a coil of wire.

    - **Diodes**: A triangle pointing to a line, with the triangle representing the anode and the line representing the cathode.

    - **Transistors**: Symbols vary for different types (NPN, PNP, FET), but they generally include three terminals (base, collector, emitter for bipolar junction transistors).

2. **Complex Components**

    - **Integrated Circuits (ICs)**: Represented by a rectangle with pins labeled to indicate the function of each pin.

- **Operational Amplifiers**: Typically shown as a triangle with input and output terminals labeled.
- **Relays**: A coil symbol with switch contacts, indicating how the relay operates.

3. **Connections and Wires**
   - **Lines**: Represent electrical connections. When two lines intersect, a dot indicates a connection, while a simple crossing without a dot indicates no connection.
   - **Nodes**: Points where components connect. They are often indicated by dots or circles.

4. **Power Sources**
   - **Batteries**: Represented by a series of alternating long and short lines.
   - **Ground**: Shown as a set of descending lines, representing the reference point for voltages in the circuit.

Reading Schematics

1. **Understanding the Flow**
   - **Left to Right**: Schematics often show the flow of the circuit from left to right, with inputs on the left and outputs on the right.
   - **Top to Bottom**: Power supplies are typically shown at the top and ground connections at the bottom.

2. **Component Relationships**
   - **Series Connections**: Components connected end-to-end with a single path for current flow.
   - **Parallel Connections**: Components connected across the same two points, providing multiple paths for current flow.

3. **Functional Blocks**
   - **Modular Approach**: Complex circuits are often divided into functional blocks, each performing a specific task (e.g., amplifier, filter, power supply).
   - **Labeling**: Blocks are usually labeled to indicate their function, making it easier to understand the overall operation of the circuit.

Practical Tips for Schematic Reading

1. **Familiarize with Symbols**
   - **Practice**: Spend time familiarizing yourself with common symbols and their variations. Practice reading schematics from simple to more complex designs.

- **Reference Guides**: Keep a reference guide or a cheat sheet of common symbols handy.

2. **Analyze Step-by-Step**
    - **Break Down**: Break down the schematic into smaller sections and analyze each section individually.
    - **Trace Paths**: Trace the path of current from the power source through the components to understand the circuit's operation.

3. **Use Color Coding**
    - **Highlighting**: Use color coding to highlight different paths or sections in the schematic. This can help visualize the flow and identify key areas.

4. **Refer to Datasheets**
    - **Component Details**: Refer to datasheets for detailed information on component specifications and pin configurations.

5. **Cross-Reference with Layouts**
    - **PCB Layouts**: Compare the schematic with the PCB (printed circuit board) layout to ensure accuracy and understanding of physical connections.

Common Schematic Reading Mistakes

1. **Ignoring Power and Ground Connections**
    - **Completeness**: Ensure all power and ground connections are accounted for. Missing these can lead to incomplete or non-functional circuits.

2. **Misinterpreting Symbols**
    - **Clarity**: Double-check symbol meanings and ensure correct interpretation. Misreading symbols can lead to incorrect assembly.

3. **Overlooking Notes and Labels**
    - **Annotations**: Pay attention to notes and labels on the schematic. They often provide crucial information about component values, configurations, and special instructions.

4. **Assuming Connections**
    - **Explicit Dots**: Ensure connections are explicitly indicated with dots. Do not assume intersecting lines are connected without confirmation.

Practice Exercises

1. **Simple Circuits**

- **Resistor-Capacitor (RC) Circuit**: Analyze a basic RC circuit to understand how resistors and capacitors interact.

- **LED Circuit**: Read and build a simple LED circuit to understand current limiting and LED operation.

2. **Intermediate Circuits**

    - **Amplifier Circuit**: Study a basic amplifier circuit, focusing on transistor operation and signal amplification.

    - **Power Supply Circuit**: Analyze a power supply schematic, including transformer, rectifier, filter, and regulator stages.

3. **Advanced Circuits**

    - **Transceiver Circuit**: Examine a ham radio transceiver schematic to understand the signal flow from antenna to speaker.

    - **Digital Logic Circuit**: Study a digital logic circuit, focusing on gates, flip-flops, and timing diagrams.

Practical Circuit Design

Practical circuit design brings together all the concepts you've learned, from basic component understanding to schematic reading and circuit analysis. This section will guide you through the steps of designing circuits that are both functional and efficient, tailored for various applications in ham radio and beyond.

Steps in Circuit Design

1. **Define the Objectives**

    - **Purpose**: Clearly define what the circuit is supposed to accomplish. For example, it could be an amplifier, a filter, or a power supply.

    - **Specifications**: List the specifications, including input and output requirements, voltage levels, current capacity, frequency response, and any special conditions (e.g., temperature range).

2. **Preliminary Design and Simulation**

    - **Block Diagram**: Create a high-level block diagram that outlines the main sections of the circuit and their interactions.

        - **Image**: Include a sample block diagram showing different sections like power supply, amplifier, filter, and output stage.

- **Component Selection**: Choose components that meet the specifications. Consider parameters like tolerance, power rating, and availability.
- **Simulation**: Use software tools like SPICE to simulate the circuit. This helps to verify the design and make adjustments before building the actual circuit.

3. **Detailed Schematic Design**
   - **Schematic Creation**: Draw a detailed schematic diagram using CAD software. Ensure that all components are correctly labeled and connected.
     - **Image**: Include an example schematic of a simple circuit, such as a low-pass filter or a basic amplifier.
   - **Verification**: Double-check the schematic for accuracy, paying attention to connections, power supply lines, and ground paths.

4. **Prototype and Testing**
   - **Breadboarding**: Build a prototype on a breadboard to test the circuit's functionality. This allows for easy modifications and troubleshooting.
     - **Image**: Show a photo or diagram of a breadboard setup with labeled components.
   - **Testing**: Use multimeters, oscilloscopes, and signal generators to test the circuit. Verify that it meets the specifications and make any necessary adjustments.
     - **Image**: Include a screenshot or photo of an oscilloscope displaying a test signal.

5. **Final Design and PCB Layout**
   - **PCB Design**: Once the prototype is verified, design a printed circuit board (PCB) layout. Use PCB design software to create the layout, ensuring proper component placement and routing.
     - **Image**: Provide a sample PCB layout with key components and traces highlighted.
   - **Fabrication**: Send the PCB design to a manufacturer for fabrication. Alternatively, you can create the PCB at home using DIY methods.
   - **Assembly**: Assemble the final circuit on the PCB, soldering all components in place.

6. **Final Testing and Documentation**
   - **Comprehensive Testing**: Perform a thorough test of the final circuit, ensuring it meets all specifications under various conditions.

- **Documentation**: Document the design process, including schematics, PCB layout, test results, and any modifications made. This documentation is crucial for future reference and troubleshooting.

Design Considerations

1. **Component Ratings**

    - **Voltage and Current Ratings**: Ensure components can handle the maximum voltage and current expected in the circuit.
    - **Power Dissipation**: Consider the power dissipation of resistors and other components to prevent overheating.

2. **Thermal Management**

    - **Heat Sinks**: Use heat sinks for components that generate significant heat, such as power transistors and voltage regulators.
    - **Ventilation**: Ensure proper ventilation in the circuit's enclosure to allow heat dissipation.

3. **Signal Integrity**

    - **Noise Reduction**: Minimize noise and interference by using proper grounding techniques and shielding sensitive components.
    - **Decoupling Capacitors**: Place decoupling capacitors near power supply pins of ICs to filter out noise and stabilize the supply voltage.

4. **Power Supply Design**

    - **Regulation**: Ensure the power supply provides a stable voltage under varying load conditions. Use voltage regulators where necessary.
    - **Filtering**: Include capacitors to filter out ripple and noise from the power supply.

5. **Safety Considerations**

    - **Overcurrent Protection**: Use fuses or circuit breakers to protect against overcurrent conditions.
    - **Insulation**: Ensure proper insulation of high-voltage sections to prevent accidental contact.

Example Project: Building a Simple RF Amplifier

1. **Objective**: Design a radio frequency (RF) amplifier to boost weak signals for better reception.
2. **Specifications**:

- **Frequency Range**: 1 MHz to 30 MHz
- **Gain**: 20 dB
- **Input/Output Impedance**: 50 ohms
- **Power Supply**: 12V DC

3. **Block Diagram**:
   - **Input Matching Network**: Matches the impedance of the source to the amplifier input.
   - **Amplifier Stage**: Provides the required gain.
   - **Output Matching Network**: Matches the amplifier output impedance to the load.
   - **Image**: Include a block diagram showing the input matching network, amplifier stage, and output matching network.

4. **Component Selection**:
   - **Transistor**: Choose a high-frequency transistor, such as the 2N2222.
   - **Resistors and Capacitors**: Select values to set the biasing and frequency response.
   - **Inductors**: Use appropriate inductors for the matching networks.

5. **Schematic Design**:
   - Draw the schematic, including all components and connections.
     - **Image**: Include a detailed schematic of the RF amplifier circuit.
   - Simulate the circuit to verify the frequency response and gain.

6. **Prototype and Testing**:
   - Build the prototype on a breadboard.
   - Test the amplifier with an RF signal generator and oscilloscope to measure gain and frequency response.
     - **Image**: Show an oscilloscope screenshot with the amplified signal waveform.

7. **PCB Design and Assembly**:
   - Design the PCB layout, ensuring short signal paths and proper grounding.
     - **Image**: Provide a PCB layout for the RF amplifier.
   - Fabricate the PCB and assemble the components.

- Perform final testing to ensure the amplifier meets the specifications.

8. **Documentation**:

    - Document the entire design process, including the schematic, PCB layout, test results, and any modifications made.

CHAPTER 5 – Licensing and Regulations Transmission Lines and Antennas

Transmission Line Basics

Transmission lines are essential components in ham radio systems, serving as the conduits that carry radio frequency (RF) signals from the transmitter to the antenna and from the antenna to the receiver. Understanding the fundamentals of transmission lines is crucial for optimizing signal transmission and ensuring efficient operation of your ham radio setup.

What Are Transmission Lines?

A transmission line is a specialized cable or other structure designed to conduct RF signals with minimal loss. Unlike regular electrical wires, transmission lines are optimized to handle high-frequency signals and maintain signal integrity over long distances.

Key Concepts

1. **Characteristic Impedance (Z_o)**

    - **Definition**: Characteristic impedance is the impedance that a transmission line presents to a signal at any point along its length. It is determined by the line's geometry and the materials used.

    - **Importance**: Matching the characteristic impedance of the transmission line to the source and load impedances is crucial for minimizing signal reflections and ensuring maximum power transfer.

2. **Velocity Factor**

    - **Definition**: The velocity factor of a transmission line is the ratio of the speed of the signal in the line to the speed of light in a vacuum. It depends on the dielectric material used in the transmission line.

    - **Typical Values**: Common velocity factors range from 0.66 to 0.95, depending on the line's construction.

3. **Losses in Transmission Lines**

    - **Types of Losses**: Transmission lines can suffer from various losses, including resistive (ohmic) losses, dielectric losses, and radiation losses.
    - **Minimizing Losses**: Using high-quality materials and proper installation techniques can help minimize losses. For example, using low-loss coaxial cables with solid shielding can reduce signal degradation.

4. **Standing Wave Ratio (SWR)**

    - **Definition**: SWR is a measure of the efficiency of power transfer from the transmission line to the load (antenna). It is the ratio of the maximum to the minimum voltage along the line.
    - **Ideal SWR**: An SWR of 1:1 indicates perfect impedance matching, while higher values indicate mismatched conditions that can lead to signal reflections and power loss.

5. **Reflection Coefficient**

    - **Definition**: The reflection coefficient measures the proportion of the signal that is reflected back due to impedance mismatching. It ranges from 0 (no reflection) to 1 (total reflection).
    - **Calculation**: $\Gamma = Z_L + Z_o / Z_L - Z_o$, where Z_L is the load impedance and Z_o is the characteristic impedance.

Types of Transmission Lines

Transmission lines come in various types, each with specific characteristics and applications. The choice of transmission line depends on factors such as frequency range, power levels, and environmental conditions.

1. **Coaxial Cable**

    - **Construction**: Consists of a central conductor, an insulating layer, a metallic shield, and an outer insulating layer.
    - **Advantages**: Offers good shielding against electromagnetic interference (EMI) and is easy to install.
    - **Applications**: Widely used in ham radio for connecting antennas to transceivers and other equipment.

2. **Twin-lead**

    - **Construction**: Consists of two parallel conductors separated by a dielectric material.
    - **Advantages**: Lower loss compared to coaxial cable at lower frequencies and is less expensive.

- **Applications**: Used for feeding antennas, especially in HF (high frequency) applications.

3. **Waveguides**
    - **Construction**: Hollow metallic structures that guide electromagnetic waves.
    - **Advantages**: Very low loss at microwave frequencies.
    - **Applications**: Used in high-frequency and high-power applications, such as radar and satellite communications.

4. **Microstrip Lines**
    - **Construction**: Consists of a conducting strip separated from a ground plane by a dielectric layer.
    - **Advantages**: Easy to fabricate and integrate with printed circuit boards (PCBs).
    - **Applications**: Used in RF and microwave circuits on PCBs.

Choosing the Right Transmission Line

Selecting the appropriate transmission line for your ham radio setup involves considering several factors:

1. **Frequency Range**: Ensure the transmission line supports the frequency range of your application.
2. **Power Handling**: Check the maximum power rating to prevent overheating and damage.
3. **Loss Characteristics**: Choose a line with low losses to maintain signal strength.
4. **Environmental Conditions**: Consider the operating environment, including temperature, humidity, and exposure to elements.

Practical Tips

1. **Installation**: Avoid sharp bends and kinks in transmission lines to prevent damage and signal loss. Use proper connectors and ensure secure connections.
2. **Maintenance**: Regularly inspect transmission lines for wear, corrosion, and damage. Replace any compromised sections to maintain optimal performance.
3. **Testing**: Use tools like SWR meters and network analyzers to test the performance of your transmission lines and identify any issues.

Types of Transmission Lines

Transmission lines are crucial in ham radio systems, providing the means to carry RF signals from one point to another with minimal loss and distortion. Different types of transmission lines are used depending on the application, frequency, and environmental conditions. This section explores the various types of transmission lines commonly used in ham radio and their specific characteristics.

Coaxial Cable

1. **Construction**

    - **Central Conductor**: Typically made of copper or aluminum, which carries the signal.

    - **Dielectric Insulator**: Surrounds the central conductor, providing insulation and maintaining the spacing between the central conductor and the outer shield.

    - **Metallic Shield**: Usually a braided or solid metal layer that surrounds the dielectric insulator, providing shielding against electromagnetic interference (EMI).

    - **Outer Insulator**: The outermost layer that protects the cable from physical damage and environmental factors.

2. **Characteristics**

    - **Impedance**: Commonly available in 50 ohms (used in RF applications) and 75 ohms (used in video and CATV applications).

    - **Attenuation**: Increases with frequency, so low-loss versions are preferred for high-frequency applications.

    - **Flexibility**: Coaxial cables are flexible and relatively easy to install.

3. **Applications**

    - Used for connecting antennas to transceivers.

    - Commonly used in feed lines for ham radio stations.

    - Ideal for use in environments with high EMI.

Twin-Lead

1. **Construction**

    - **Parallel Conductors**: Two parallel wires separated by a consistent distance using an insulating material.

    - **Dielectric Spacer**: Often made of plastic, it maintains the spacing and provides structural support.

2. **Characteristics**

    - **Impedance**: Typically 300 ohms.

    - **Losses**: Lower loss compared to coaxial cables at lower frequencies due to less dielectric material.

    - **Susceptibility to Interference**: More susceptible to EMI, requiring careful routing away from metal objects and other potential sources of interference.

3. **Applications**

    - Used in HF (high frequency) antenna systems.

    - Common in dipole antenna feed lines.

Waveguides

1. **Construction**

    - **Hollow Metal Tubes**: Usually rectangular or circular, designed to guide electromagnetic waves.

    - **Material**: Typically made from metal like copper or aluminum to provide a conductive path for the RF signals.

2. **Characteristics**

    - **Frequency Range**: Very low loss at microwave frequencies, unsuitable for lower frequencies.

    - **Power Handling**: Capable of handling high power levels with minimal loss.

3. **Applications**

    - Used in radar systems, satellite communications, and other high-frequency applications.

    - Ideal for transmitting signals over long distances without significant loss.

Microstrip Lines

1. **Construction**

    - **Conducting Strip**: A thin metallic strip on a dielectric substrate.

    - **Ground Plane**: A conductive layer on the opposite side of the substrate from the conducting strip.

2. **Characteristics**

- **Integration**: Easily integrated into printed circuit boards (PCBs), making it suitable for compact RF and microwave circuits.
- **Impedance Control**: Allows precise control of characteristic impedance through design parameters.

3. **Applications**
   - Used in RF and microwave circuits on PCBs.
   - Common in mobile devices and other compact electronics.

Comparative Table of Transmission Lines

Type	Construction	Impedance	Frequency Range	Applications
Coaxial Cable	Central conductor, dielectric insulator, metallic shield, outer insulator	50 ohms (RF), 75 ohms (video/CATV)	Low to high frequencies	Connecting antennas, feed lines in ham radio
Twin-Lead	Parallel conductors with dielectric spacer	300 ohms	Low to medium frequencies	HF antenna systems, dipole feed lines
Waveguides	Hollow metal tubes	Varies	High to very high frequencies	Radar, satellite communications
Microstrip Lines	Conducting strip on dielectric substrate with ground plane	Varies	Medium to high frequencies	RF and microwave circuits on PCBs

Choosing the Right Transmission Line

Selecting the appropriate transmission line involves understanding the specific requirements of your application:

1. **Frequency**: Ensure the transmission line supports the frequency range of your application without excessive loss.
2. **Power Handling**: Check the maximum power rating to prevent overheating and damage.
3. **Environmental Factors**: Consider the operating environment, including exposure to elements and potential for physical wear and tear.
4. **Impedance Matching**: Match the characteristic impedance of the transmission line with the source and load to minimize reflections and losses.

Practical Tips

1. **Installation and Routing**: Avoid sharp bends and physical stress on the transmission lines to prevent damage and maintain performance.
2. **Regular Inspection**: Periodically inspect transmission lines for signs of wear, corrosion, and damage, especially in outdoor installations.
3. **Testing and Maintenance**: Use SWR meters and network analyzers to regularly test and ensure the integrity and performance of the transmission lines.

Understanding the types of transmission lines and their appropriate applications is crucial for efficient signal transmission in ham radio setups. In the next section, we will explore impedance matching, a vital concept for optimizing power transfer between components in your radio system.

Impedance Matching

Impedance matching is a fundamental concept in ham radio and electronic circuit design. It's all about ensuring that the impedance of different components within your system aligns perfectly. This alignment is critical for maximizing power transfer and minimizing signal reflections. Proper impedance matching can significantly enhance the performance and efficiency of your radio communications, ensuring you get the most out of your equipment.

Why Impedance Matching Matters

Think of impedance matching as the secret sauce that makes your radio setup work seamlessly. When the impedance of your transmission line matches that of both the source and the load, magic happens: you get maximum power transfer with minimal signal reflections. This is crucial because mismatched impedances can lead to power loss, signal distortion, and even potential damage to your transmitter.

Imagine you have a high-performance sports car, but the tires are not inflated correctly. No matter how powerful the engine is, you won't get the best performance. Similarly, in your radio setup, even if you have top-notch equipment, mismatched impedance can hinder performance.

1. **Maximizing Power Transfer**
    - **Theory**: The maximum power transfer theorem states that power transfer to a load is maximized when the load impedance matches the source impedance. In practical terms, for most ham radio equipment, this means ensuring your antenna impedance matches the transmitter's output impedance, usually 50 ohms.
2. **Minimizing Reflections**
    - **SWR (Standing Wave Ratio)**: SWR is a measure of impedance matching quality. An SWR of 1:1 indicates perfect matching, while higher values indicate increasing levels of mismatch. High SWR can result in reflected signals, causing interference and reducing the effective radiated power, which translates to weaker signals and shorter communication range.

Techniques for Achieving Impedance Matching

There are several methods to achieve impedance matching, each with its own set of advantages and applications. Here are some of the most common techniques:

1. **Quarter-Wave Transformer**

    - **Principle**: A transmission line section that is one-quarter wavelength long can transform impedances. The characteristic impedance of this section should be the geometric mean of the source and load impedances.

    - **Application**: Commonly used in VHF and UHF applications, where the physical length of the quarter-wave section is manageable. For instance, if you need to match a 50-ohm transmitter to a 75-ohm antenna, a quarter-wave transformer with an impedance of about 61 ohms can do the job.

2. **Baluns (Balanced to Unbalanced Transformers)**

    - **Function**: Baluns are used to connect balanced devices (like dipole antennas) to unbalanced devices (like coaxial cables), helping with impedance matching.

    - **Types**: Common types include 1:1 and 4:1 baluns, which transform impedance by a factor of 1 or 4, respectively. These are essential in antenna systems to ensure proper impedance matching and minimize RF interference.

3. **Stub Matching**

    - **Principle**: Uses short sections of transmission line (stubs) connected either in parallel or series with the main line to adjust the impedance.

    - **Application**: Often used in HF and VHF bands where precise matching is required. Stubs can be cut to specific lengths to provide the necessary impedance transformation. For example, adding a parallel stub to a mismatched feed line can help achieve better matching.

4. **LC Networks (Inductance and Capacitance)**

    - **Types**: Include L-networks, T-networks, and Pi-networks, which use combinations of inductors and capacitors to achieve impedance matching.

    - **Flexibility**: These networks are highly versatile and can match a wide range of impedances, making them ideal for applications where the frequency range varies significantly.

    - **Design Considerations**: The choice of inductance and capacitance values depends on the desired impedance transformation and the operating frequency. For example, designing an L-network to match a 50-ohm transmitter to a 300-ohm antenna involves selecting appropriate inductance and capacitance values to achieve the transformation.

Practical Examples

1. **Matching an Antenna to a Transmitter**

    - **Scenario**: You have a transmitter with an output impedance of 50 ohms and an antenna with an impedance of 75 ohms.

    - **Solution**: Using a quarter-wave transformer with a characteristic impedance of $\sqrt{50 \times 75} \approx 61$ ohms effectively matches the transmitter to the antenna, ensuring efficient power transfer and minimizing reflections.

2. **Using a Balun for a Dipole Antenna**

    - **Scenario**: You're connecting a balanced dipole antenna to an unbalanced 50-ohm coaxial cable.

    - **Solution**: A 1:1 balun transitions from the balanced antenna to the unbalanced cable, ensuring proper impedance matching and minimizing RF interference, thus optimizing your setup.

3. **Stub Matching for HF Bands**

    - **Scenario**: An antenna system operating at 14 MHz with a slight impedance mismatch.

    - **Solution**: Adding a parallel stub (a short section of transmission line) to the feed line can adjust the impedance and achieve a better match, improving signal quality and transmission efficiency.

Tools for Impedance Matching

1. **SWR Meter**

    - **Function**: Measures the standing wave ratio in your transmission line, helping to identify the level of impedance mismatch.

    - **Usage**: Place the SWR meter between the transmitter and the antenna system to monitor SWR while making adjustments. This ensures you can see the impact of your changes in real time.

2. **Antenna Analyzer**

    - **Function**: Provides a detailed analysis of the antenna system, including impedance, SWR, and resonance points.

    - **Usage**: An antenna analyzer offers a comprehensive check of your antenna system, making it easier to diagnose and fix impedance matching issues. This tool is particularly useful for fine-tuning and ensuring optimal performance.

3. **Network Analyzer**

    - **Function**: Offers advanced measurements of network parameters, including impedance, reflection coefficient, and S-parameters.

- **Usage**: Typically used in more complex systems and for precise tuning of components in RF circuits. Network analyzers are invaluable for detailed analysis and troubleshooting.

Antenna Theory and Design

Antenna theory and design are central to ham radio operations. Antennas are the critical interface between the radio and the airwaves, determining the efficiency and range of your communications. Understanding how antennas work and learning how to design them can significantly enhance your ham radio experience.

Basic Antenna Concepts

1. **Radiation Pattern**

    - **Definition**: The radiation pattern of an antenna describes how it radiates energy into space. It is typically represented in a polar plot, showing the relative strength of the signal in different directions.

    - **Types**: Common radiation patterns include omnidirectional (radiating equally in all directions) and directional (focusing energy in specific directions).

2. **Gain**

    - **Definition**: Antenna gain measures how well an antenna directs radio frequency energy compared to an isotropic radiator (a theoretical antenna that radiates equally in all directions).

    - **Units**: Gain is measured in decibels (dB). Higher gain indicates a more focused radiation pattern, enhancing the signal strength in desired directions.

3. **Polarization**

    - **Definition**: Polarization refers to the orientation of the electric field of the radio wave radiated by the antenna. Common types include vertical, horizontal, and circular polarization.

    - **Importance**: Matching the polarization of the transmitting and receiving antennas maximizes signal reception and reduces losses.

4. **Bandwidth**

    - **Definition**: The range of frequencies over which an antenna operates effectively. It is usually expressed as a percentage of the center frequency or in absolute frequency units (MHz).

- **Consideration**: Antennas with a wide bandwidth can operate across multiple frequencies, making them versatile for different applications.

5. **Impedance**
    - **Definition**: The impedance of an antenna is the ratio of voltage to current at its terminals. It is typically 50 ohms for most ham radio antennas to match the standard transmission line impedance.
    - **Matching**: Ensuring the antenna impedance matches the transmission line and transmitter impedance minimizes reflections and maximizes power transfer.

Types of Antennas

1. **Dipole Antenna**
    - **Description**: A simple and effective antenna consisting of two conductive elements (usually metal wires) that are half a wavelength long.
    - **Radiation Pattern**: Typically has a figure-eight pattern in the plane perpendicular to the antenna.
    - **Applications**: Widely used for HF bands due to its simplicity and effectiveness.

2. **Yagi-Uda Antenna**
    - **Description**: A directional antenna consisting of a driven element, reflector, and one or more directors. The elements are arranged parallel to each other on a boom.
    - **Gain and Directivity**: Offers high gain and good directivity, making it ideal for long-distance communications and TV reception.
    - **Applications**: Commonly used in VHF and UHF bands.

3. **Vertical Antenna**
    - **Description**: A monopole antenna, typically a quarter wavelength long, mounted vertically. It uses the ground or a ground plane as the counterpoise.
    - **Radiation Pattern**: Omnidirectional in the horizontal plane, making it suitable for ground wave and low-angle skywave propagation.
    - **Applications**: Popular for mobile and base station use in HF and VHF bands.

4. **Loop Antenna**
    - **Description**: Consists of a loop (or multiple loops) of wire, often in a circular or square shape. Can be small (magnetic loop) or large (full-wave loop).

- **Radiation Pattern**: Depends on the loop size and shape, offering various patterns from omnidirectional to directional.

- **Applications**: Effective for HF bands and used in both transmitting and receiving applications.

5. **Log-Periodic Dipole Array (LPDA)**

    - **Description**: A broadband antenna consisting of multiple dipole elements of varying lengths, arranged in a log-periodic pattern.

    - **Frequency Range**: Capable of operating over a wide range of frequencies with consistent performance.

    - **Applications**: Used in HF and VHF bands for wideband communication and TV reception.

Designing an Antenna

Designing an antenna involves understanding the desired operating frequency, gain, radiation pattern, and physical constraints. Here's a step-by-step approach:

1. **Define Requirements**

    - **Frequency Range**: Determine the operating frequency or frequencies for your antenna.

    - **Gain and Directivity**: Decide on the required gain and whether the antenna needs to be directional or omnidirectional.

    - **Physical Constraints**: Consider the available space and mounting options.

2. **Choose Antenna Type**

    - Based on the requirements, select an appropriate antenna type (e.g., dipole, Yagi-Uda, vertical).

3. **Calculate Dimensions**

    - **Dipole Example**: For a half-wave dipole, the length L can be calculated using the formula $L=f/468$ feet, where f is the frequency in MHz.

    - **Yagi-Uda Example**: Determine the lengths and spacing of the elements based on the design formulas and desired frequency.

4. **Simulate and Optimize**

    - Use antenna design software (e.g., EZNEC, 4NEC2) to simulate the antenna performance. Adjust dimensions and element spacing to optimize the gain, radiation pattern, and impedance.

5. **Construct the Antenna**
   - Build the antenna using suitable materials (e.g., aluminum tubing for elements, insulators for mounting points).
   - Ensure mechanical stability and durability, especially for outdoor installations.

6. **Test and Adjust**
   - Measure the SWR and impedance using an antenna analyzer. Make adjustments as necessary to achieve the desired performance.
   - Fine-tune the antenna by adjusting element lengths or spacing based on the test results.

Practical Considerations

1. **Materials**
   - Use weather-resistant materials for outdoor antennas. Aluminum is common for elements due to its light weight and good conductivity.
   - Ensure all connections are corrosion-resistant, especially in coastal or humid environments.

2. **Mounting and Safety**
   - Mount antennas securely to withstand wind and weather. Use guy wires for stability if necessary.
   - Follow safety guidelines to avoid contact with power lines and ensure proper grounding to protect against lightning strikes.

3. **Grounding and Lightning Protection**
   - Proper grounding is essential for safety and performance. Use ground rods and heavy-gauge wire to connect the antenna system to the earth.
   - Install lightning arrestors and surge protectors to safeguard equipment from lightning-induced surges.

Example: Designing a Simple Dipole Antenna

Let's walk through the process of designing a basic half-wave dipole antenna for the 20-meter band (14 MHz):

1. **Define Requirements**
   - **Frequency**: 14 MHz
   - **Gain**: Omnidirectional in the horizontal plane

- **Physical Constraints**: Must fit in a backyard space

2. **Calculate Dimensions**

    - **Length**: $L=468/14 \approx 33.4$ feet total, or about 16.7 feet for each leg of the dipole.

3. **Simulate and Optimize**

    - Use an antenna simulation tool to check the radiation pattern and impedance. Adjust lengths if needed.

4. **Construct the Antenna**

    - Cut two wires to 16.7 feet each.
    - Connect the wires to a center insulator and feed them with a 50-ohm coaxial cable through a balun.

5. **Test and Adjust**

    - Measure the SWR using an antenna analyzer. Trim the wire lengths slightly if the SWR is not optimal.

Types of Antennas

Choosing the right antenna is essential for effective ham radio communication. Different types of antennas are optimized for various frequencies, propagation characteristics, and installation environments. Understanding the strengths and weaknesses of each type can help you select the best antenna for your needs.

Dipole Antenna

1. **Description**

    - The dipole antenna is one of the simplest and most efficient antennas. It consists of two conductive elements (usually metal wires) that are each a quarter wavelength long.
    - **Configuration**: The classic configuration is the half-wave dipole, where each leg is a quarter wavelength, making the total length half a wavelength.

2. **Radiation Pattern**

    - **Horizontal Dipole**: Exhibits a figure-eight pattern in the plane perpendicular to the wire, providing good broadside coverage.
    - **Vertical Dipole**: Radiates omnidirectionally in the horizontal plane, suitable for ground wave and low-angle skywave propagation.

3. **Applications**

- Ideal for HF bands, offering good performance with simple construction.
- Commonly used in both fixed and portable setups due to its ease of deployment.

4. **Advantages and Disadvantages**
    - **Advantages**: Simple design, easy to build, and effective performance.
    - **Disadvantages**: Requires space for the length of the wire and may need support structures for higher frequencies.

Yagi-Uda Antenna

1. **Description**
    - The Yagi-Uda antenna, commonly known as a Yagi antenna, consists of a driven element (usually a half-wave dipole), a reflector, and one or more directors.
    - **Configuration**: The elements are mounted parallel to each other on a boom, with the reflector behind the driven element and the directors in front.

2. **Radiation Pattern**
    - Highly directional with a main lobe pointing in the direction of the directors.
    - Provides significant gain and directivity, making it excellent for long-distance communication.

3. **Applications**
    - Common in VHF and UHF bands for applications like TV reception, satellite communication, and DXing (long-distance communication).

4. **Advantages and Disadvantages**
    - **Advantages**: High gain, directional, and can be customized for specific frequencies.
    - **Disadvantages**: More complex design and construction, requires precise alignment, and larger physical size.

Vertical Antenna

1. **Description**
    - A vertical antenna is a monopole antenna, typically a quarter wavelength long, mounted vertically. It uses the ground or a ground plane as the counterpoise.
    - **Configuration**: Often includes radials or a ground plane to enhance performance.

2. **Radiation Pattern**

- Omnidirectional in the horizontal plane, making it suitable for ground wave propagation.
- Radiates at low angles, which is ideal for long-distance (DX) communication, especially in HF bands.

3. **Applications**
   - Popular for mobile and base station use, particularly in HF and VHF bands.
   - Ideal for limited space environments where horizontal antennas are impractical.

4. **Advantages and Disadvantages**
   - **Advantages**: Omnidirectional pattern, requires less horizontal space, and easy to install.
   - **Disadvantages**: Needs a good ground system or radials, and can be affected by nearby structures.

Loop Antenna

1. **Description**
   - Loop antennas consist of a loop (or multiple loops) of wire, often in a circular or square shape. They can be small (magnetic loop) or large (full-wave loop).
   - **Configuration**: Can be resonant (tuned to a specific frequency) or broadband (covering a range of frequencies).

2. **Radiation Pattern**
   - Varies with size and shape, but typically offers good omnidirectional or slightly directional patterns.
   - Magnetic loops are highly efficient and have a small footprint, making them ideal for limited space.

3. **Applications**
   - Effective for HF bands, often used in both transmitting and receiving applications.
   - Magnetic loops are popular for portable and stealth installations due to their compact size.

4. **Advantages and Disadvantages**
   - **Advantages**: Compact size, good performance in limited space, and reduced noise pickup.

- **Disadvantages**: Can be complex to construct and tune, especially for resonant designs.

Log-Periodic Dipole Array (LPDA)

1. **Description**
    - The LPDA is a broadband antenna consisting of multiple dipole elements of varying lengths, arranged in a log-periodic pattern.
    - **Configuration**: Elements are connected in a specific sequence to provide wideband performance.

2. **Radiation Pattern**
    - Directional with moderate gain, providing consistent performance over a wide frequency range.
    - Offers a broader bandwidth compared to Yagi antennas.

3. **Applications**
    - Used in HF and VHF bands for wideband communication, including TV and radio reception.
    - Ideal for applications requiring consistent performance across multiple frequencies.

4. **Advantages and Disadvantages**
    - **Advantages**: Wide bandwidth, moderate gain, and relatively simple to construct for wideband use.
    - **Disadvantages**: Larger physical size compared to single-frequency antennas, and somewhat lower gain than a Yagi for the same size.

Comparative Table of Antennas

Antenna Type	Description	Radiation Pattern	Applications	Advantages	Disadvantages
Dipole	Simple two-element design	Figure-eight or omnidirectional	HF bands, fixed and portable setups	Simple, easy to build, effective	Requires space, needs support structures
Yagi-Uda	Driven element with reflector and directors	Highly directional	VHF, UHF, long-distance communication	High gain, customizable, directional	Complex design, requires alignment
Vertical	Monopole with ground plane	Omnidirectional	HF, VHF, mobile and base stations	Omnidirectional, less space needed	Needs ground system, affected by surroundings
Loop	Circular or square wire loop	Omnidirectional or directional	HF bands, transmitting and receiving	Compact, good in limited space, reduced noise	Complex construction, tuning required
Log-Periodic Dipole Array (LPDA)	Multiple dipoles in log-periodic pattern	Directional with wide bandwidth	HF, VHF, wideband communication	Wide bandwidth, moderate gain, simple	Larger size, lower gain than Yagi

Practical Considerations

1. **Material Selection**

    - Use durable materials like aluminum or copper for elements to ensure longevity and reliability, especially for outdoor installations.

    - Ensure all connections are corrosion-resistant, particularly in humid or coastal environments.

2. **Installation**

    - Securely mount antennas to withstand environmental factors like wind and weather. Use guy wires for additional stability if necessary.

    - Ensure proper grounding and lightning protection to safeguard your equipment and enhance performance.

3. **Testing and Maintenance**

    - Regularly inspect antennas for wear and tear. Check connections and replace any corroded or damaged parts.

    - Use tools like SWR meters and antenna analyzers to test performance and make necessary adjustments to maintain optimal operation.

Antenna Installation and Safety

Installing an antenna properly is crucial for optimal performance and safety. Whether you are setting up a new antenna system or upgrading an existing one, understanding the best practices for installation and safety can help you avoid common pitfalls and ensure reliable operation.

Site Selection

1. **Location**

    - **Elevation**: Higher is generally better. Placing your antenna on a rooftop or a tower can improve signal reception and transmission by reducing obstructions.

    - **Clearance**: Ensure there is adequate clearance from trees, buildings, and other structures that can obstruct the signal path.

    - **Accessibility**: Choose a location that is easily accessible for installation and maintenance but safe from unauthorized access.

2. **Orientation**

    - **Directionality**: For directional antennas like Yagi-Uda, point the antenna in the direction of your most desired contacts or coverage area.

    - **Avoiding Interference**: Position your antenna away from sources of electromagnetic interference (EMI), such as power lines, transformers, and other electronic devices.

Mounting the Antenna

1. **Support Structures**

    - **Masts and Towers**: Use sturdy masts or towers designed to withstand environmental stressors like wind and precipitation.

    - **Guy Wires**: For tall structures, guy wires can provide additional stability. Ensure they are properly anchored and tensioned.

2. **Mounting Hardware**

    - **Brackets and Clamps**: Use high-quality, weather-resistant brackets and clamps to secure the antenna to the support structure.

    - **Insulators**: For wire antennas, insulators can help prevent unwanted grounding and maintain the integrity of the antenna system.

3. **Safety Considerations**

- **Avoid Power Lines**: Always keep a safe distance from power lines to prevent electrical hazards. Use the "10-foot rule" – stay at least 10 feet away from power lines at all times.

- **Weather Precautions**: Install antennas in calm weather conditions to reduce the risk of accidents. Avoid working on antennas during storms or high winds.

Grounding

1. **Purpose of Grounding**

   - **Lightning Protection**: Proper grounding helps protect your equipment and home from lightning strikes. It provides a low-resistance path to dissipate the energy safely into the ground.

   - **Static Discharge**: Grounding helps reduce the buildup of static electricity, which can cause interference and damage sensitive equipment.

2. **Grounding Methods**

   - **Ground Rods**: Install ground rods made of copper or galvanized steel into the earth near the antenna site. Connect the antenna system to these rods using heavy-gauge wire.

   - **Ground Wire**: Use a short, straight run of heavy-gauge wire to connect the antenna to the ground rod. Avoid sharp bends and loops in the wire to minimize resistance.

   - **Bonding**: Bond all ground connections together to create a single grounding system. This includes the antenna, tower, and any nearby metal structures.

3. **Lightning Arrestors**

   - **Installation**: Install lightning arrestors on the coaxial feed line where it enters your home or shack. These devices help protect your equipment by diverting the surge from a lightning strike to ground.

   - **Maintenance**: Regularly inspect and replace lightning arrestors as needed to ensure they remain effective.

Safety Protocols

1. **Personal Safety**

   - **Protective Gear**: Wear appropriate safety gear, including gloves, hard hats, and non-slip shoes, when installing or maintaining antennas.

   - **Ladders and Harnesses**: Use stable ladders and safety harnesses when working at heights. Ensure all climbing equipment is in good condition and properly secured.

2. **Electrical Safety**

- **Turn Off Power**: Before working on any part of your antenna system, ensure all associated equipment is powered off to prevent electric shock.
- **Check for Voltage**: Use a voltage tester to confirm there is no electrical potential present on any part of the antenna system before touching it.

3. **Environmental Considerations**
   - **Weather Awareness**: Be aware of weather conditions. Avoid working on antennas during adverse weather, such as thunderstorms or high winds.
   - **Wildlife and Vegetation**: Check for wildlife habitats and overgrown vegetation around the antenna site. Ensure your activities do not disturb local wildlife and trim any vegetation that could interfere with the antenna.

Maintenance and Inspection

1. **Routine Checks**
   - **Visual Inspection**: Regularly inspect your antenna system for signs of wear and tear, such as corrosion, loose connections, and physical damage.
   - **Connection Tightness**: Ensure all bolts, clamps, and connectors are tight and secure. Loose connections can degrade performance and pose safety risks.

2. **Performance Testing**
   - **SWR Meter**: Use an SWR meter to regularly check the standing wave ratio of your antenna system. An increase in SWR can indicate issues such as damage or detuning.
   - **Antenna Analyzer**: An antenna analyzer provides a detailed assessment of your antenna's performance, including impedance and resonance. Use it to fine-tune and optimize your setup.

3. **Repairs and Upgrades**
   - **Component Replacement**: Replace any damaged or degraded components, such as corroded connectors or frayed wires, to maintain optimal performance.
   - **System Upgrades**: Periodically review your antenna system and consider upgrades to improve performance, such as adding additional elements to a Yagi-Uda antenna or installing a higher-quality coaxial cable.

Practical Tips

1. **Labeling and Documentation**
   - **Label Connections**: Clearly label all connections and components in your antenna system. This helps with troubleshooting and future maintenance.

- **Keep Records**: Maintain a log of all installation and maintenance activities, including dates, observations, and any changes made to the system. This documentation can be invaluable for diagnosing issues and planning future upgrades.

2. **Community Resources**

    - **Local Clubs**: Join local ham radio clubs to share knowledge and experiences with other enthusiasts. Club members can often provide valuable insights and assistance with antenna installation and maintenance.

    - **Online Forums**: Participate in online forums and discussion groups dedicated to ham radio. These platforms can be great sources of information and support.

Grounding and Lightning Protection

Grounding and lightning protection are crucial aspects of setting up and maintaining a ham radio station. Proper grounding not only protects your equipment from lightning strikes but also helps reduce noise and improve signal quality. Let's dive into the essentials of grounding and how to safeguard your setup from lightning damage.

Importance of Grounding

Imagine your ham radio station as a finely tuned orchestra. Each instrument (or component) needs to play in harmony for the best performance. Grounding is like the conductor, ensuring that everything works together smoothly. It provides a reference point for all voltages in your system, reduces noise, and offers a path for unwanted electrical energy to safely dissipate into the earth.

Think of grounding as the unsung hero of your station, quietly doing its job behind the scenes. Without proper grounding, you might experience interference, signal degradation, or even equipment failure. More critically, improper grounding can pose serious safety risks, especially during thunderstorms.

Setting Up a Grounding System

1. **Ground Rods**

    - Ground rods are metal rods driven into the earth near your station. They serve as the primary pathway for electrical currents to safely dissipate.

    - When installing ground rods, aim for a depth of at least eight feet. This ensures a solid connection with the earth, providing an effective grounding point. Copper-clad steel rods are a popular choice due to their durability and conductivity.

2. **Connecting Your Equipment**

    - Use heavy-gauge wire to connect your equipment to the ground rods. The thicker the wire, the lower its resistance, which is crucial for handling the high currents associated with lightning strikes.

- Keep these connections as short and straight as possible. Avoid sharp bends or loops, as these can increase resistance and reduce the effectiveness of your grounding system.

3. **Bonding Ground Systems**

    - If you have multiple ground rods or grounding points, ensure they are all bonded together. This creates a single, unified grounding system, minimizing potential differences between different parts of your setup.

    - Use heavy-duty clamps and conductive materials to make these connections, ensuring they are secure and resistant to corrosion.

Lightning Protection

While grounding is essential for day-to-day operation, lightning protection is about preparing for those rare but potentially devastating events when Mother Nature shows her fury. A well-designed lightning protection system can mean the difference between minor inconvenience and catastrophic equipment loss.

1. **Lightning Arrestors**

    - Lightning arrestors are devices that divert excess voltage from lightning strikes away from your equipment. Install these on all cables entering your home or shack.

    - Place arrestors on your coaxial feed lines where they enter your station. These devices will shunt the high-voltage surge to ground before it can reach your sensitive equipment.

2. **Surge Protectors**

    - Use surge protectors on all electrical outlets connected to your ham radio equipment. These protect against transient voltage spikes that can occur during electrical storms.

    - Choose surge protectors with a high joule rating for maximum protection. Ensure they are properly grounded to your station's grounding system.

3. **Antenna Disconnection**

    - During severe thunderstorms, it's a good idea to physically disconnect your antennas from your equipment. This simple step can provide an additional layer of protection.

    - Use quick-disconnect fittings to make this process easier. Store the disconnected ends away from your equipment to prevent accidental contact.

Practical Grounding and Lightning Protection Tips

1. **Regular Inspections**

- Periodically inspect all grounding connections for signs of corrosion or loosening. Over time, exposure to the elements can degrade these connections, reducing their effectiveness.
- Check ground rods and wires for integrity. Replace any components that show signs of wear or damage.

2. **Maintaining a Low Resistance Path**

- Ensure the ground around your rods remains moist, as dry soil can significantly increase resistance. Watering the area around your ground rods during dry periods can help maintain good conductivity.
- If possible, use multiple ground rods spaced several feet apart and connected together. This can improve overall grounding effectiveness, especially in poor soil conditions.

3. **Documentation and Testing**

- Keep a detailed log of your grounding and lightning protection setup, including installation dates, materials used, and any maintenance performed. This can be invaluable for troubleshooting and future upgrades.
- Test your grounding system periodically with a ground resistance tester. Aim for a ground resistance of less than 5 ohms to ensure effective protection.

Real-Life Scenario: A Ham Radio Operator's Experience

Consider the story of Dave, a ham radio operator from Florida, a state known for its frequent thunderstorms. Dave's meticulous approach to grounding and lightning protection saved his station during a particularly severe storm.

Dave had installed multiple ground rods around his property, connected with heavy-gauge copper wire. He used high-quality lightning arrestors on all his coaxial lines and surge protectors on his power outlets. During the storm, lightning struck a tree near his house, and the surge traveled through the ground.

Thanks to his well-designed grounding system, the electrical surge was safely dissipated. The lightning arrestors and surge protectors did their job, preventing any damage to his radio equipment. Dave's careful preparation paid off, allowing him to continue his ham radio activities without interruption.

Community Resources and Support

Don't hesitate to seek advice from other ham radio operators in your local club or online forums. Many experienced hams are willing to share their knowledge and provide tips based on their own experiences with grounding and lightning protection. Joining a community can offer valuable insights and help you troubleshoot any issues you encounter.

By implementing these grounding and lightning protection strategies, you can safeguard your ham radio station against potential threats. Proper grounding not only enhances your station's performance but also protects your investment and ensures your safety. Taking the time to set up an effective grounding system is a small effort that can provide significant peace of mind.

CHAPTER 6 – Radio Equipment

Transceivers and Receivers

Transceivers and receivers are at the heart of any ham radio station. Understanding their functions, types, and how to select the right equipment can significantly enhance your ham radio experience.

What is a Transceiver?

A transceiver is a device that combines both a transmitter and a receiver in a single unit. This integration allows for more efficient operation and ease of use, as both functions share common components like the power supply, antenna system, and frequency control. Transceivers are popular because they save space and make it easier to switch between transmitting and receiving without the need for separate equipment.

Example: The Icom IC-7300 is a popular transceiver among ham radio operators. It features an integrated design that provides both transmission and reception capabilities, making it versatile for various operating conditions.

What is a Receiver?

A receiver, on the other hand, is a device specifically designed to pick up and process incoming radio signals. Receivers are critical for listening to and decoding transmissions from other operators. High-quality receivers can enhance your ability to pick up weak signals and reduce noise and interference.

Example: The Kenwood R-600 is a dedicated receiver known for its sensitivity and selectivity, making it ideal for receiving distant or faint signals.

Types and Functions

Different types of transceivers and receivers cater to various needs and operating conditions. Understanding their functions can help you choose the right equipment for your ham radio setup.

Types of Transceivers

1. **Base Station Transceivers**
    - **Description**: These are larger units designed for fixed installation in a ham radio station. They typically offer more features and higher power outputs.

- **Example**: The Yaesu FT-991A is a comprehensive base station transceiver that supports multiple bands and modes, including HF, VHF, and UHF.

2. **Mobile Transceivers**

    - **Description**: Designed for installation in vehicles, these transceivers are compact and rugged. They usually operate on VHF and UHF bands, which are well-suited for mobile communication.

    - **Example**: The Icom IC-2730A is a dual-band mobile transceiver known for its durability and ease of use in mobile environments.

3. **Handheld Transceivers (HTs)**

    - **Description**: Also known as "handy-talkies," these are portable transceivers that allow for communication on the go. They are battery-powered and typically operate on VHF and UHF bands.

    - **Example**: The Baofeng UV-5R is a popular handheld transceiver due to its affordability and versatility.

Types of Receivers

1. **Communications Receivers**

    - **Description**: These are high-performance receivers designed for serious monitoring of HF and VHF/UHF bands. They offer advanced features like multiple bandwidth settings, digital signal processing, and superior selectivity.

    - **Example**: The Drake R8B is a well-regarded communications receiver with excellent sensitivity and selectivity, suitable for DXing and general listening.

2. **Software-Defined Receivers (SDRs)**

    - **Description**: SDRs use software to process radio signals, providing flexibility and the ability to upgrade features through software updates. They are often used with a computer interface.

    - **Example**: The FlexRadio FLEX-6500 is a powerful SDR that offers advanced signal processing and a wide range of customizable features.

3. **Scanning Receivers**

    - **Description**: These receivers can automatically scan multiple frequencies to find active signals. They are popular for monitoring public service communications, air traffic control, and other non-amateur radio services.

    - **Example**: The Uniden Bearcat BC125AT is a widely used scanning receiver with a broad frequency range and programmable channels.

Selecting the Right Equipment

Choosing the right transceiver or receiver depends on several factors, including your operating preferences, budget, and the specific requirements of your ham radio activities.

Factors to Consider

1. **Frequency Coverage**

   - **Needs Assessment**: Determine which frequency bands you need to cover. HF bands are essential for long-distance communication, while VHF and UHF bands are useful for local and regional contacts.

   - **Example**: If you are interested in DXing on the HF bands, a transceiver with comprehensive HF coverage is crucial.

2. **Power Output**

   - **Legal Limits**: Ensure the transceiver complies with legal power output limits for your license class and operating region.

   - **Typical Requirements**: For most amateur activities, a transceiver with 100 watts output is sufficient. For mobile and handheld transceivers, power outputs are typically lower.

3. **Modes of Operation**

   - **Modes**: Consider the modes you plan to operate, such as CW (Morse code), SSB (single sideband), AM, FM, and digital modes like FT8 or PSK31.

   - **Versatility**: Choose equipment that supports multiple modes if you plan to experiment with different types of communication.

4. **Ease of Use**

   - **User Interface**: Look for a transceiver or receiver with an intuitive user interface and easy-to-read display.

   - **Learning Curve**: Some advanced features may require a steeper learning curve, so consider your level of expertise and willingness to learn.

5. **Budget**

   - **Initial Investment**: Determine your budget for the initial purchase and any additional accessories needed, such as antennas, power supplies, and microphones.

   - **Long-Term Value**: Consider the long-term value of the equipment, including potential upgrades and maintenance costs.

Practical Tips

1. **Try Before You Buy**

    - **Hamfests and Club Meetings**: Attend hamfests, club meetings, and other events where you can see and try different equipment. This hands-on experience can help you make an informed decision.

    - **Reviews and Recommendations**: Read reviews and ask for recommendations from experienced operators. Online forums and social media groups can also be valuable resources.

2. **Consider Used Equipment**

    - **Quality and Cost**: High-quality used equipment can offer significant savings. Ensure the equipment has been well-maintained and is in good working condition.

    - **Reputable Sellers**: Buy from reputable sellers or through trusted platforms to avoid potential issues with used gear.

Amplifiers and Power Supplies

Amplifiers and power supplies are crucial components that enhance the performance and reliability of your ham radio station. Understanding their functions and types can help you optimize your setup.

Amplifiers

Amplifiers boost the power output of your transceiver, allowing you to transmit stronger signals and reach distant stations more effectively. There are different types of amplifiers, each suited to specific applications.

1. **Linear Amplifiers**

    - **Description**: Linear amplifiers increase the amplitude of the RF signal without significantly distorting it. They are essential for modes like SSB and CW, where signal purity is crucial.

    - **Example**: The Ameritron AL-811H is a popular linear amplifier that provides reliable power amplification for HF bands.

2. **Solid-State Amplifiers**

    - **Description**: These amplifiers use solid-state devices like transistors for amplification. They are known for their efficiency, reliability, and compact size.

    - **Example**: The Elecraft KPA500 is a well-regarded solid-state amplifier that offers high efficiency and ease of use.

3. **Tube Amplifiers**

- **Description**: Tube amplifiers use vacuum tubes for amplification. They are valued for their ability to handle high power levels and robustness.
- **Example**: The Heathkit SB-220 is a classic tube amplifier known for its durability and performance.

Power Supplies

Power supplies provide the necessary electrical power for your transceiver and other equipment. Selecting the right power supply ensures stable and reliable operation.

1. **Linear Power Supplies**
    - **Description**: Linear power supplies provide a steady and noise-free DC output, making them ideal for sensitive radio equipment.
    - **Example**: The Astron RS-35A is a widely used linear power supply known for its reliability and clean output.

2. **Switching Power Supplies**
    - **Description**: Switching power supplies are more efficient and compact than linear power supplies. They convert AC to DC using high-frequency switching techniques.
    - **Example**: The Samlex SEC-1235M is a popular switching power supply that offers high efficiency and lightweight design.

Understanding Amplifier Types

Amplifiers are categorized based on their design and functionality. Understanding these types can help you choose the right amplifier for your needs.

Class A Amplifiers

1. **Description**
    - **Operation**: Class A amplifiers conduct signal over the entire 360-degree cycle of the input waveform, ensuring high fidelity and low distortion.
    - **Efficiency**: They are less efficient compared to other classes, often converting a significant amount of power into heat.

2. **Applications**
    - **Usage**: Suitable for applications where signal purity is critical, such as audio and low-power RF amplification.

Class B Amplifiers

1. **Description**

- **Operation**: Class B amplifiers conduct signal for 180 degrees of the input waveform cycle. They require two devices (one for each half of the cycle) to produce a full output signal.
- **Efficiency**: More efficient than Class A amplifiers but can introduce distortion at the crossover point of the waveform.

2. **Applications**
   - **Usage**: Often used in push-pull configurations to improve linearity and reduce distortion, suitable for higher power applications.

Class AB Amplifiers

1. **Description**
   - **Operation**: Class AB amplifiers combine the principles of Class A and Class B, conducting signal for more than 180 degrees but less than 360 degrees of the input cycle.
   - **Efficiency**: They offer a compromise between the low distortion of Class A and the higher efficiency of Class B.

2. **Applications**
   - **Usage**: Widely used in RF amplification where both efficiency and linearity are important.

Class C Amplifiers

1. **Description**
   - **Operation**: Class C amplifiers conduct for less than 180 degrees of the input waveform cycle, making them highly efficient but non-linear.
   - **Efficiency**: They are very efficient, converting most of the input power into output RF power, with little waste heat.

2. **Applications**
   - **Usage**: Ideal for applications where signal distortion can be tolerated, such as in frequency modulation (FM) and continuous wave (CW) transmission.

Practical Considerations for Amplifiers

1. **Heat Management**
   - **Cooling Systems**: Ensure amplifiers have adequate cooling systems, such as fans or heatsinks, to dissipate heat and prevent overheating.

- **Ventilation**: Install equipment in well-ventilated areas to enhance airflow and cooling efficiency.

2. **Power Requirements**
    - **Current Capacity**: Ensure your power supply can provide sufficient current for all connected devices, including peak power demands.
    - **Voltage Stability**: Choose a power supply with stable voltage output to prevent fluctuations that could affect your equipment's performance.

3. **Interference and Noise**
    - **Filtering**: Use power supplies with good filtering capabilities to minimize electrical noise that could interfere with your signals.
    - **Grounding**: Properly ground all equipment to reduce noise and ensure safety.

Practical Example: Setting Up a Base Station Transceiver

Let's say you're setting up a base station transceiver in your ham shack. You've chosen the Yaesu FT-991A for its versatility and power. Here's how you would go about setting up your new equipment:

1. **Unpacking and Initial Setup**
    - **Placement**: Find a suitable spot for your transceiver. It should be on a stable surface with adequate ventilation to keep it cool.
    - **Connections**: Connect the power supply to your transceiver. If you're using the Astron RS-35A linear power supply, ensure the connections are secure and the voltage settings are correct.

2. **Antenna Connection**
    - **Coaxial Cable**: Use a high-quality coaxial cable to connect your transceiver to your antenna system. Double-check that the connectors are properly attached and free from corrosion.
    - **SWR Check**: Use an SWR meter to check the standing wave ratio of your antenna setup. Aim for an SWR of 1:1 for optimal performance.

3. **Power On and Basic Configuration**
    - **Power Up**: Turn on the power supply and then the transceiver. Listen for any unusual noises that might indicate a problem.
    - **Frequency Selection**: Use the tuning knob to select the desired frequency band. The Yaesu FT-991A supports a wide range of frequencies, so you can experiment with different bands.

4. **Initial Testing**

    - **Transmit and Receive**: Test both transmitting and receiving functions. Make a few test calls to check the clarity and strength of your signal.

    - **Adjust Settings**: Fine-tune the settings such as microphone gain, RF power output, and audio levels to suit your preferences and ensure clear communication.

Tips for Long-Term Maintenance

1. **Regular Cleaning**

    - **Dust Removal**: Keep your equipment dust-free. Dust can accumulate inside vents and affect cooling.

    - **Connection Checks**: Periodically check all connections to ensure they remain tight and free from corrosion.

2. **Software Updates**

    - **Firmware Updates**: Check the manufacturer's website for firmware updates. These can improve performance and add new features to your transceiver.

    - **Backup Settings**: Regularly back up your transceiver settings, especially before performing updates.

3. **Routine Testing**

    - **Performance Monitoring**: Use an antenna analyzer and SWR meter to regularly test your antenna system's performance. Make adjustments as needed to maintain optimal efficiency.

    - **Power Supply Check**: Ensure your power supply continues to deliver stable voltage and current. Replace any aging or faulty components promptly.

CHAPTER 7 – Digital Modes and Software

Introduction to Digital Modes

Digital modes in ham radio represent a fascinating and rapidly evolving aspect of the hobby. These modes use digital signals to encode information, allowing for more efficient and reliable communication. They are particularly useful in conditions where traditional voice communication might struggle due to poor propagation or high noise levels. Let's explore some of the most popular digital modes and understand why they are favored by many amateur radio operators.

What Are Digital Modes?

Digital modes refer to methods of communication that use digital signals instead of analog. These modes often involve converting text or data into digital signals that can be transmitted over radio frequencies and then decoded back into text or data at the receiving end. This process typically involves a computer or dedicated digital mode device connected to a transceiver.

Example: Digital modes like FT8 and PSK31 can enable communication over long distances with relatively low power and minimal bandwidth, making them ideal for DXing and weak signal work.

Popular Digital Modes

1. **FT8 (Franke-Taylor 8)**

    - **Description**: FT8 is a relatively new digital mode designed for making reliable, confirmed contacts under extreme weak-signal conditions. It uses 8-FSK modulation and is known for its efficiency and robustness.

    - **Usage**: FT8 has quickly become one of the most popular digital modes in the ham radio community. It is particularly favored for its ability to decode signals well below the noise floor, making it perfect for DXing and operating in challenging conditions.

    - **Example**: FT8 signals are typically short, 15-second transmissions, allowing for rapid contact exchanges even when signal strengths are very low.

2. **PSK31 (Phase Shift Keying 31)**

    - **Description**: PSK31 is an older but still widely used digital mode. It uses phase shift keying to encode data, allowing for narrow bandwidth transmissions. PSK31 is noted for its simplicity and ease of use.

    - **Usage**: PSK31 is popular for casual conversation (ragchewing) and for making contacts when conditions are less than ideal. Its narrow bandwidth helps it punch through noise and interference.

    - **Example**: A typical PSK31 conversation might involve exchanging names, locations, and signal reports, with the text displayed in real-time on the computer screen.

3. **RTTY (Radio Teletype)**

    - **Description**: RTTY is one of the oldest digital modes still in use. It uses frequency shift keying (FSK) to transmit text, originally developed for mechanical teleprinters.

- **Usage**: Despite its age, RTTY remains popular for contesting and DXing. Its robust nature and relatively simple equipment requirements keep it relevant in modern ham radio.

- **Example**: During a RTTY contest, operators can rapidly exchange call signs, signal reports, and other information, often using automated software to speed up the process.

Software for Ham Radio

Modern ham radio operations often rely heavily on software to enhance communication capabilities and streamline station management. From logging contacts to decoding digital signals, software plays a critical role in the contemporary ham radio experience.

Logging Software

Logging software is essential for keeping track of contacts, managing QSL cards, and participating in awards and contests. Good logging software helps organize and maintain records efficiently, ensuring that all relevant information is readily accessible.

1. **Features to Look For**

    - **Ease of Use**: The software should have an intuitive interface, making it easy to log contacts quickly and accurately.

    - **Integration**: Look for software that integrates well with other tools and equipment, such as digital mode software, online QSL services, and contesting software.

    - **Customization**: The ability to customize fields and views can help tailor the logging software to your specific needs and preferences.

2. **Popular Logging Software**

    - **Ham Radio Deluxe (HRD)**

        - **Description**: HRD is a comprehensive suite that includes logging, digital mode operation, and rig control. It is highly regarded for its wide range of features and user-friendly interface.

        - **Usage**: HRD can handle all aspects of station management, from logging contacts and managing awards to controlling your radio and operating digital modes.

    - **N1MM Logger+**

        - **Description**: N1MM Logger+ is a free contest logging software that is very popular among contesters. It offers extensive support for various contests and integrates with many digital mode applications.

- **Usage**: Contest operators use N1MM Logger+ to streamline the logging process during high-intensity contests, allowing for quick and accurate entry of contact information.

- **Logbook of The World (LoTW)**
    - **Description**: LoTW is a secure, online contact verification system operated by the ARRL. It allows hams to upload logs and receive credit for awards.
    - **Usage**: Many operators use LoTW in conjunction with their logging software to easily manage and verify contacts for awards like DXCC and WAS.

Software for Digital Modes

Digital mode software is essential for encoding and decoding digital signals. These programs work in conjunction with your transceiver and computer to facilitate digital communication.

1. **WSJT-X**

    - **Description**: WSJT-X is a popular software suite developed by Joe Taylor, K1JT. It supports various digital modes, including FT8, FT4, JT65, and more. WSJT-X is particularly well-known for its weak-signal capabilities.

    - **Usage**: Operators use WSJT-X to make contacts under difficult conditions, leveraging its ability to decode signals well below the noise floor. The software provides a user-friendly interface for managing contacts and adjusting settings.

    - **Example**: During a typical FT8 session, WSJT-X handles the encoding and decoding of signals, automating the exchange of call signs, grid locators, and signal reports.

2. **FLDIGI**

    - **Description**: FLDIGI is a versatile digital mode software that supports a wide range of modes, including PSK31, RTTY, and Olivia. It is known for its flexibility and ease of use.

    - **Usage**: FLDIGI is often used for casual digital mode operation and experimentation. It offers extensive customization options, allowing operators to fine-tune their setup for optimal performance.

    - **Example**: An operator using FLDIGI for a PSK31 QSO can easily monitor the waterfall display, type messages, and adjust settings to improve signal quality.

3. **MMSSTV**

    - **Description**: MMSSTV is specialized software for Slow Scan Television (SSTV), a mode used to transmit images over radio frequencies. It is popular among hams who enjoy visual communication.

- **Usage**: Operators use MMSSTV to send and receive images, often engaging in SSTV contests or simply sharing pictures with other hams around the world.

- **Example**: During an SSTV transmission, MMSSTV encodes an image into audio tones that are transmitted via the radio. The receiving station decodes these tones back into the image.

Practical Examples and Tips

Let's look at a practical example of how you might set up and use digital mode software in your ham radio station. Suppose you want to start using FT8 with WSJT-X.

1. **Setting Up WSJT-X**

    - **Download and Install**: First, download WSJT-X from the official website and install it on your computer.

    - **Configure Radio Settings**: Open WSJT-X and configure the radio settings to match your transceiver's model and connection type (e.g., USB, serial port).

    - **Audio Settings**: Ensure your computer's audio settings are configured correctly, allowing WSJT-X to receive and transmit audio signals through your transceiver.

2. **Making Contacts with FT8**

    - **Select Mode**: In WSJT-X, select FT8 from the mode menu.

    - **Choose Frequency**: Tune your transceiver to an FT8 frequency, typically found in the HF bands.

    - **Start Monitoring**: Begin monitoring the band for FT8 signals. WSJT-X will display decoded signals in the main window.

    - **Call CQ**: If you see an open frequency, you can call CQ by clicking the "Enable Tx" button. WSJT-X will handle the transmission sequence automatically.

    - **Logging Contacts**: Once a contact is made, log the contact details in your logging software, or use WSJT-X's built-in logging feature.

3. **Tips for Success**

    - **Timing**: FT8 operates on precise time synchronization. Ensure your computer's clock is accurate, using time synchronization software if necessary.

    - **Signal Reports**: Pay attention to signal reports to gauge your signal's reach and adjust power levels or antenna settings as needed.

    - **Practice and Patience**: Digital modes can have a learning curve. Practice regularly and be patient as you learn the nuances of each mode.

Example Table: Popular Digital Modes Comparison

Mode	Bandwidth	Typical Usage	Advantages	Disadvantages
FT8	50 Hz	DXing, Weak Signals	Excellent weak-signal performance	Limited to short exchanges
PSK31	31.25 Hz	Ragchewing, Casual	Narrow bandwidth, easy to use	Susceptible to noise
RTTY	170 Hz	Contesting, DXing	Robust and well-supported	Requires good signal quality

Integrating Software with Your Station

Integrating digital mode software and logging software with your station setup can significantly enhance your ham radio experience. Here's how you can streamline the process:

1. **Connecting Your Transceiver**

    - **Interface Devices**: Use interface devices such as Signalink USB or RigExpert to connect your transceiver to your computer. These devices handle audio and keying connections, making the setup straightforward.

    - **CAT Control**: Utilize CAT (Computer-Aided Transceiver) control for seamless operation. This allows your software to control the transceiver's frequency, mode, and other settings directly.

2. **Automating Processes**

    - **Macros**: Use macros in digital mode software to automate repetitive tasks, such as sending standard messages. This can speed up operations during contests or regular QSOs.

    - **Logging Integration**: Set up your logging software to automatically log contacts made through digital modes. Many programs offer seamless integration, reducing the risk of manual entry errors.

3. **Remote Operation**

    - **Remote Access Software**: Use remote access software like TeamViewer or AnyDesk to operate your station remotely. This setup allows you to manage and operate your station from anywhere with an internet connection.

    - **Cloud Logging**: Consider cloud-based logging solutions that enable you to access your logs from any device. This is particularly useful for operators who travel frequently.

Digital modes and the associated software have transformed the landscape of ham radio. They offer new ways to communicate, especially in challenging conditions, and open up a world of possibilities for experimentation and learning. Whether you're chasing distant DX stations with FT8, enjoying a casual chat on PSK31, or competing in a RTTY contest, digital modes provide a rich and rewarding aspect of amateur radio.

By carefully selecting and integrating digital mode software and logging tools, you can enhance your operating experience, streamline your workflows, and make the most of the technological advancements available to modern ham radio operators.

CHAPTER 8 – Propagation and Space Weather

Understanding Propagation

Propagation refers to the way radio waves travel through the atmosphere from the transmitter to the receiver. Understanding the different modes of propagation is essential for maximizing the effectiveness of your ham radio communications. Various factors influence how radio waves propagate, including frequency, time of day, atmospheric conditions, and geographic location.

Key Concepts of Propagation

1. **Line-of-Sight Propagation**

    - **Description**: This mode occurs when radio waves travel directly from the transmitter to the receiver without any obstacles. It is typical for VHF and UHF frequencies, where the waves generally do not bend or reflect significantly.

    - **Limitations**: The range is limited to the visual horizon, which can be extended slightly by the curvature of the Earth and the height of the antennas.

2. **Ground Wave Propagation**

    - **Description**: In this mode, radio waves travel along the Earth's surface. It is most effective at lower frequencies, particularly in the LF and MF bands.

    - **Applications**: Ground wave propagation is used for AM radio broadcasts and maritime communications. It can provide reliable communication over medium distances, especially during the day.

3. **Skywave Propagation**

    - **Description**: Skywave propagation occurs when radio waves are reflected or refracted back to Earth by the ionosphere. This mode allows for long-distance communication, often thousands of kilometers.

- **Frequency Range**: Typically used in the HF bands (3-30 MHz), where ionospheric reflection is most effective.
- **Dependence on the Ionosphere**: The effectiveness of skywave propagation varies with ionospheric conditions, which are influenced by solar activity and time of day.

4. **Sporadic E Propagation**
    - **Description**: This phenomenon occurs when signals are reflected by dense, irregular patches of ionization in the E layer of the ionosphere. It is unpredictable but can support long-distance VHF communication.
    - **Frequency Range**: Commonly affects frequencies between 10 MHz and 200 MHz, making it useful for VHF DXing during the sporadic E season.

5. **Tropospheric Propagation**
    - **Description**: Tropospheric propagation involves the bending (refraction) of radio waves in the lower atmosphere (troposphere), often enhancing VHF and UHF signals over long distances.
    - **Weather Influence**: Conditions such as temperature inversions, high pressure systems, and weather fronts can create ducts that trap radio waves, allowing them to travel beyond the normal line-of-sight range.

Practical Considerations for Propagation

1. **Frequency Selection**
    - **HF Bands**: For long-distance communication, especially at night, HF bands are typically the best choice due to effective skywave propagation.
    - **VHF/UHF Bands**: For local communication and when using repeaters, VHF and UHF bands are preferred due to their line-of-sight propagation characteristics.

2. **Antenna Design**
    - **Height**: Raising the antenna height can improve both line-of-sight and ground wave propagation.
    - **Type**: Directional antennas (e.g., Yagi) can enhance signal strength in a particular direction, improving both skywave and tropospheric propagation.

3. **Monitoring Propagation Conditions**
    - **Beacons**: Listening to propagation beacons can provide real-time information about current propagation conditions.
    - **Online Tools**: Websites and apps that track solar activity and ionospheric conditions can help predict when certain types of propagation will be most effective.

Ionospheric Layers

The ionosphere plays a crucial role in radio wave propagation, particularly for HF communications. It consists of several layers of ionized gases that reflect or refract radio waves back to Earth, enabling long-distance communication. Understanding the ionospheric layers helps in predicting how radio waves will behave and when certain frequencies will be most effective.

Structure of the Ionosphere

1. **D Layer**

    - **Characteristics**: The D layer is the lowest ionospheric layer, located approximately 60 to 90 kilometers above the Earth's surface. It is primarily present during daylight hours due to solar radiation.

    - **Impact on Propagation**: The D layer absorbs low-frequency (LF and MF) radio waves, causing significant attenuation. This absorption decreases at night, allowing these frequencies to travel further.

2. **E Layer**

    - **Characteristics**: Situated between 90 and 120 kilometers above the Earth, the E layer is also influenced by solar radiation. It is most notable for its role in sporadic E propagation.

    - **Impact on Propagation**: The E layer can reflect higher frequency HF signals (up to about 10 MHz) during the day. Sporadic E events can also support VHF propagation, making it possible to achieve long-distance contacts on bands typically limited to line-of-sight.

3. **F Layer**

    - **Characteristics**: The F layer is divided into two sub-layers, F1 and F2, during the daytime. At night, these layers merge into a single F layer. The F2 layer, located around 200 to 400 kilometers above the Earth, is the most significant for HF propagation.

    - **Impact on Propagation**: The F2 layer can reflect radio waves back to Earth, allowing for global communication on HF bands. It is influenced by solar activity, with higher levels of ionization during peak solar periods.

Diurnal Variations

1. **Daytime Conditions**

    - **Increased Ionization**: During the day, increased solar radiation enhances the ionization of the D, E, and F layers. This increased ionization supports higher frequency propagation but can also lead to greater absorption of lower frequencies by the D layer.

- **Frequency Considerations**: Higher HF frequencies (e.g., 15-30 MHz) are more effective during the day as they can be reflected by the highly ionized F2 layer.

2. **Nighttime Conditions**

    - **Reduced Ionization**: At night, the lack of solar radiation causes the D layer to dissipate, reducing absorption of lower frequencies. The E layer also weakens, while the F layers merge and retain enough ionization to reflect HF signals.

    - **Frequency Considerations**: Lower HF frequencies (e.g., 3.5-7 MHz) become more effective at night, as they can travel longer distances with reduced attenuation.

Seasonal Variations

1. **Summer**

    - **Higher Ionization**: Longer daylight hours and more direct solar radiation increase ionization levels, particularly in the D and E layers. This can enhance sporadic E propagation and support higher frequencies.

    - **Challenges**: Increased D layer absorption can limit low-frequency propagation during the day.

2. **Winter**

    - **Lower Ionization**: Shorter days and lower solar angles reduce ionization levels, particularly in the D layer. This reduction in absorption benefits low-frequency propagation.

    - **Advantages**: HF propagation, particularly on lower frequencies, can be excellent during the winter months due to reduced D layer absorption.

Solar Activity and the Ionosphere

Solar activity, such as sunspots, solar flares, and coronal mass ejections (CMEs), has a significant impact on the ionosphere and, consequently, on radio wave propagation.

1. **Sunspots**

    - **Description**: Sunspots are temporary phenomena on the Sun's surface that appear as dark spots due to lower temperatures. They are associated with increased solar activity.

    - **Impact on Propagation**: High sunspot numbers correlate with increased ionization in the F2 layer, enhancing HF propagation.

2. **Solar Flares and CMEs**

    - **Description**: Solar flares are sudden bursts of radiation from the Sun's surface, while CMEs are massive bursts of solar wind and magnetic fields.

- **Impact on Propagation**: These events can cause short-term disruptions in HF propagation (radio blackouts) and increase auroral activity, affecting VHF propagation.

Day and Night Propagation

Radio wave propagation varies significantly between day and night due to changes in the ionosphere's structure and ionization levels. Understanding these differences is crucial for effective communication, as it helps operators choose the right frequencies and optimize their antenna setups.

Daytime Propagation

1. **Increased Ionization**

    - **Solar Radiation**: During the day, solar radiation increases the ionization levels in the ionospheric layers, particularly the D, E, and F layers. This enhanced ionization supports the propagation of higher frequency signals.

    - **Higher Frequencies**: Frequencies in the higher HF range (15-30 MHz) benefit from daytime ionization, as they can be reflected by the highly ionized F2 layer, enabling long-distance communication.

2. **D Layer Absorption**

    - **Low-Frequency Attenuation**: The D layer, being the lowest ionospheric layer, absorbs lower frequency signals (below 10 MHz) during the day. This absorption can significantly attenuate these signals, reducing their effective range.

    - **HF Propagation**: Despite this absorption, the HF bands can still be effective for long-distance communication, especially on frequencies that can penetrate the D layer and be reflected by the F2 layer.

3. **Sporadic E Propagation**

    - **E Layer Patches**: Sporadic E propagation occurs when patches of dense ionization form in the E layer. These patches can reflect VHF signals, allowing for long-distance contacts that are not usually possible on these frequencies.

    - **Seasonal Variations**: Sporadic E events are more common during the summer months, enhancing VHF propagation and providing opportunities for DXing on bands like 6 meters.

Nighttime Propagation

1. **Reduced Ionization**

    - **Absence of Solar Radiation**: At night, the lack of solar radiation leads to decreased ionization in the D and E layers. This reduction lowers the absorption of low-frequency signals, improving their range.

- **F Layer Persistence**: The F2 layer retains enough ionization to continue reflecting HF signals, though its ionization levels are lower than during the day.

2. **Low-Frequency Enhancement**

    - **Improved Range**: With the D layer dissipating at night, lower HF frequencies (3.5-7 MHz) experience less absorption and can travel longer distances. This enhancement makes the 80 and 40-meter bands particularly effective for nighttime communication.

    - **Gray Line Propagation**: The transition period between day and night, known as the gray line, can provide enhanced propagation conditions. Signals traveling along the gray line path can experience reduced attenuation and increased range.

3. **Challenges and Considerations**

    - **Noise Levels**: Nighttime propagation can be affected by increased atmospheric noise, particularly on the lower HF bands. Operators may need to use noise reduction techniques to maintain clear communications.

    - **Frequency Selection**: Choosing the right frequency for nighttime operation is crucial. Lower HF frequencies are generally more effective, but operators should be prepared to switch bands as conditions change.

Practical Tips for Day and Night Propagation

1. **Monitor Propagation Conditions**

    - **Real-Time Tools**: Use real-time propagation tools and websites to monitor current conditions. These resources provide valuable information on ionospheric activity, solar flux, and other factors influencing propagation.

    - **Beacons and Nets**: Listen to propagation beacons and participate in ham radio nets to get a sense of current band conditions and receive updates from other operators.

2. **Adjusting Antennas**

    - **Antenna Height**: Raising or lowering your antenna can impact its effectiveness for different propagation modes. Experiment with antenna height to find the optimal setup for both day and night operations.

    - **Directional Antennas**: Use directional antennas to focus your signal in the desired direction. This can enhance signal strength and improve your ability to make contacts during varying propagation conditions.

3. **Flexibility and Adaptation**

    - **Band Switching**: Be prepared to switch bands as conditions change throughout the day and night. Having a multi-band antenna or multiple antennas can facilitate this flexibility.

- **Power Adjustments**: Adjust your power output as needed. Higher power can help overcome challenging conditions, but be mindful of legal limits and the potential for causing interference

Space Weather and its Effects

Space weather refers to the environmental conditions in space as influenced by the Sun and the solar wind. These conditions can have significant effects on radio wave propagation, particularly for HF communications. Understanding space weather and its impacts can help operators predict and adapt to changes in propagation conditions.

Components of Space Weather

1. **Solar Wind**

    - **Description**: The solar wind is a stream of charged particles (plasma) released from the upper atmosphere of the Sun. It flows outward through the solar system and interacts with the Earth's magnetic field.

    - **Impact on Propagation**: Variations in the solar wind can affect the Earth's ionosphere and magnetic field, leading to changes in radio wave propagation. Increased solar wind activity can enhance ionospheric disturbances, affecting HF communication.

2. **Geomagnetic Storms**

    - **Description**: Geomagnetic storms are temporary disturbances in the Earth's magnetosphere caused by solar wind and solar flares. These storms can last from several hours to days.

    - **Impact on Propagation**: Geomagnetic storms can cause significant disruptions to HF propagation by altering the ionosphere's structure. This can lead to increased absorption of HF signals, reduced signal strength, and increased noise levels.

3. **Solar Flares**

    - **Description**: Solar flares are sudden bursts of intense radiation from the Sun's surface, often associated with sunspots. They can release a large amount of energy across the electromagnetic spectrum.

    - **Impact on Propagation**: Solar flares can cause short-term HF radio blackouts on the sunlit side of the Earth. These blackouts occur when the increased radiation temporarily ionizes the lower ionosphere (D layer), absorbing HF signals and preventing them from reaching their destinations.

4. **Coronal Mass Ejections (CMEs)**

    - **Description**: CMEs are massive bursts of solar wind and magnetic fields released into space. They can take several days to reach Earth.

- **Impact on Propagation**: When CMEs interact with the Earth's magnetosphere, they can trigger geomagnetic storms, leading to disruptions in HF propagation. CMEs can also enhance auroral activity, affecting VHF propagation in polar regions.

Monitoring Space Weather

1. **Solar Observatories**

    - **NASA and NOAA**: Organizations like NASA and NOAA (National Oceanic and Atmospheric Administration) monitor solar activity and provide real-time data on space weather conditions. These observations help predict potential impacts on radio wave propagation.

    - **Online Resources**: Websites such as SpaceWeather.com and the NOAA Space Weather Prediction Center offer up-to-date information on solar activity, including sunspot numbers, solar flare alerts, and geomagnetic storm forecasts.

2. **Propagation Prediction Tools**

    - **VOACAP**: The Voice of America Coverage Analysis Program (VOACAP) is a popular tool for predicting HF propagation conditions based on current and forecasted space weather data.

    - **Solar Data Apps**: Mobile apps and software programs that provide real-time solar data and propagation forecasts can be valuable tools for operators looking to stay informed about space weather conditions.

Practical Tips for Dealing with Space Weather

1. **Stay Informed**

    - **Regular Monitoring**: Regularly check space weather forecasts and updates from reliable sources. Staying informed about current and expected conditions can help you plan your operating schedule and frequency choices.

    - **Alerts and Notifications**: Sign up for space weather alerts and notifications to receive timely information about solar flares, CMEs, and geomagnetic storms that could affect your operations.

2. **Adapt Your Operating Practices**

    - **Frequency Adjustments**: During periods of high solar activity or geomagnetic storms, higher HF frequencies may be less affected by absorption and disturbances. Be prepared to switch frequencies to maintain reliable communication.

    - **Lower Power Operations**: In conditions where noise levels are high due to space weather, consider reducing your power output to avoid contributing to the noise floor and to comply with good operating practices.

3. **Experiment and Learn**

    - **Track Conditions**: Keep a log of propagation conditions and your operating experiences during different space weather events. This can help you identify patterns and improve your understanding of how space weather affects your specific station setup.

    - **Join Discussions**: Participate in ham radio forums, clubs, and online groups to share experiences and learn from other operators who have dealt with similar conditions

Solar Activity

Solar activity plays a critical role in shaping the conditions of the ionosphere and, consequently, the propagation of radio waves. Understanding the different aspects of solar activity can help ham radio operators anticipate changes in propagation conditions and optimize their operations.

Types of Solar Activity

1. **Sunspots**

    - **Description**: Sunspots are temporary dark spots on the Sun's surface caused by magnetic activity. They are cooler than the surrounding areas and can last from a few days to several months.

    - **Impact on Propagation**: High sunspot numbers generally indicate increased solar activity, which can enhance the ionization of the ionosphere's F2 layer. This increased ionization improves HF propagation, particularly for higher frequencies.

2. **Solar Flares**

    - **Description**: Solar flares are intense bursts of radiation caused by the sudden release of magnetic energy on the Sun's surface. They can emit energy across the electromagnetic spectrum, including X-rays and ultraviolet light.

    - **Impact on Propagation**: Solar flares can cause sudden ionospheric disturbances (SIDs), leading to short-term HF radio blackouts on the sunlit side of the Earth. These blackouts occur when the increased radiation temporarily enhances the D layer's ionization, absorbing HF signals.

3. **Coronal Mass Ejections (CMEs)**

    - **Description**: CMEs are large expulsions of plasma and magnetic fields from the Sun's corona. They can cause geomagnetic storms when they collide with the Earth's magnetosphere.

    - **Impact on Propagation**: CMEs can disrupt HF and VHF propagation by causing geomagnetic storms and increasing auroral activity. These effects can lead to signal fading, increased noise levels, and unpredictable propagation conditions.

4. **Solar Wind**

    - **Description**: The solar wind is a stream of charged particles continuously emitted by the Sun. It interacts with the Earth's magnetosphere, influencing geomagnetic conditions.

    - **Impact on Propagation**: Variations in the solar wind can affect the ionosphere and geomagnetic field, leading to changes in HF propagation. High-speed solar wind streams can enhance ionospheric disturbances and affect long-distance communication.

Solar Cycles

1. **11-Year Solar Cycle**

    - **Description**: Solar activity follows an approximately 11-year cycle, characterized by periods of high and low sunspot numbers. The cycle includes a solar maximum (peak activity) and a solar minimum (low activity).

    - **Impact on Propagation**: During the solar maximum, increased sunspot numbers enhance ionospheric ionization, improving HF propagation. Conversely, during the solar minimum, reduced solar activity can lead to poorer HF propagation conditions, particularly on higher frequencies.

2. **Long-Term Trends**

    - **Historical Patterns**: Long-term trends in solar activity can affect propagation conditions over decades. Periods of unusually high or low solar activity, such as the Maunder Minimum, have significant impacts on radio propagation.

    - **Current Cycle**: Understanding the current position within the solar cycle helps operators anticipate propagation conditions. Tools like solar cycle prediction charts and sunspot number forecasts provide valuable insights.

Monitoring Solar Activity

1. **Solar Observatories**

    - **NASA and NOAA**: Organizations like NASA and NOAA monitor solar activity and provide real-time data on sunspots, solar flares, and CMEs. These observations are critical for predicting space weather impacts on radio propagation.

    - **SOHO and SDO**: The Solar and Heliospheric Observatory (SOHO) and the Solar Dynamics Observatory (SDO) are key missions that provide detailed images and data on the Sun's activity.

2. **Online Resources**

- **Websites and Apps**: Websites like SpaceWeather.com and the NOAA Space Weather Prediction Center offer up-to-date information on solar activity. Mobile apps also provide real-time solar data and alerts.
- **Amateur Radio Tools**: Software and tools designed for ham radio operators, such as solar flux monitors and propagation prediction programs, help track solar activity and its effects on propagation.

Practical Tips for Operating During Varying Solar Activity

1. **Frequency Adjustments**
   - **High Activity**: During periods of high solar activity, higher HF frequencies (15-30 MHz) are more likely to be effective for long-distance communication due to enhanced F2 layer ionization.
   - **Low Activity**: During low solar activity, lower HF frequencies (3.5-7 MHz) may provide more reliable propagation, as higher frequencies may not be sufficiently reflected by the ionosphere.

2. **Operating Practices**
   - **Flexible Scheduling**: Plan your operating schedule around solar activity forecasts. Be prepared to adjust your frequency and mode of operation based on current conditions.
   - **Monitoring and Logging**: Keep a log of solar activity and its impact on your communications. This can help you understand patterns and improve your ability to predict favorable conditions

Predicting Propagation Conditions

Accurately predicting propagation conditions is essential for effective ham radio operation. By understanding the factors that influence propagation and using the right tools, operators can enhance their ability to make contacts and optimize their station's performance.

Key Factors Influencing Propagation

1. **Solar Activity**
   - **Sunspots and Solar Flares**: High sunspot numbers and solar flares increase ionospheric ionization, improving HF propagation. Monitoring these activities helps predict favorable conditions.
   - **CMEs and Solar Wind**: Coronal mass ejections and variations in the solar wind can cause geomagnetic storms, affecting HF and VHF propagation.

2. **Ionospheric Conditions**

- **Diurnal Variations**: Daytime ionization supports higher frequency propagation, while nighttime conditions favor lower frequencies. Understanding these variations helps in selecting the right operating times and frequencies.
- **Seasonal Changes**: Seasonal variations affect ionospheric ionization, with summer conditions enhancing sporadic E propagation and winter favoring low-frequency HF bands.

3. **Geomagnetic Activity**
   - **K-index and A-index**: The K-index measures geomagnetic activity over short periods, while the A-index provides a longer-term average. High values indicate geomagnetic storms, which can disrupt HF propagation.
   - **Auroral Activity**: Increased geomagnetic activity can enhance auroral propagation on VHF bands, providing opportunities for long-distance contacts in polar regions.

Tools for Predicting Propagation

1. **VOACAP (Voice of America Coverage Analysis Program)**
   - **Description**: VOACAP is a widely used tool for predicting HF propagation conditions. It uses a variety of inputs, including solar activity, to provide detailed forecasts.
   - **Usage**: Operators can input their location, antenna specifications, and desired frequency to receive propagation predictions for different times of day and seasons.

2. **Online Propagation Tools**
   - **Websites and Apps**: Websites like DX Summit and QRZ.com offer real-time propagation forecasts and reports from other operators. Mobile apps like HamAlert provide notifications for favorable conditions.
   - **Solar Data Sites**: SpaceWeather.com and the NOAA Space Weather Prediction Center offer solar activity updates and propagation predictions, helping operators stay informed.

3. **Beacons and Nets**
   - **Propagation Beacons**: Monitoring propagation beacons can provide real-time information about current conditions. Beacons transmit on specific frequencies and are designed to be heard over long distances.
   - **Ham Radio Nets**: Participating in ham radio nets allows operators to share real-time propagation reports and receive updates from others experiencing similar conditions.

Practical Tips for Predicting and Using Propagation

1. **Regular Monitoring**

- **Daily Checks**: Make it a habit to check propagation forecasts and solar activity reports daily. This helps you stay ahead of changing conditions and plan your operating times effectively.

- **Use Multiple Sources**: Rely on a combination of tools and resources to get a comprehensive view of propagation conditions. Different tools may offer unique insights that can enhance your predictions.

2. **Experimentation and Logging**

- **Track Conditions**: Keep a log of propagation conditions and your operating experiences. Note the times, frequencies, and any unusual propagation events. This log can help you identify patterns and improve your predictions.

- **Experiment with Frequencies**: Try different frequencies and modes during varying conditions to see what works best. Experimentation helps you understand how your station performs under different propagation scenarios.

3. **Adjusting Station Setup**

- **Antenna Adjustments**: Be prepared to adjust your antenna setup based on propagation conditions. Changing the height, orientation, or type of antenna can improve your signal strength and reach.

- **Power Output**: Adjust your power output as needed. Higher power can help overcome challenging conditions, but always operate within legal limits and be mindful of potential interference.

Example Table: Propagation Prediction Tools Comparison

Tool	Features	Pros	Cons
VOACAP	Detailed HF propagation forecasts	Accurate, customizable inputs	Requires initial setup
DX Summit	Real-time reports, operator inputs	Community-driven, real-time updates	Dependent on user activity
SpaceWeather.com	Solar activity updates, forecasts	Comprehensive space weather info	May require interpretation
HamAlert	Notifications for favorable conditions	Mobile-friendly, customizable alerts	Limited to specific alerts

CHAPTER 9 – Building and Mantaining Your Station

Station Design and Layout

Designing and laying out your ham radio station is a crucial step in creating an efficient, enjoyable, and safe operating environment. A well-thought-out station maximizes your ability to communicate effectively while minimizing potential issues related to space, ergonomics, and interference.

Planning Your Station

The foundation of a successful station begins with careful planning. Consider your operating goals, the space you have available, and your budget. Start by sketching a basic layout of your station, visualizing the placement of essential equipment such as your transceiver, power supply, computer, and any accessories. Think about how you will route cables to keep them organized and ensure all components are easily accessible.

Maximizing your available space is essential. If possible, allocate a dedicated room or a specific corner of a room for your station. This helps keep all equipment organized and reduces the risk of interference from other electronic devices. Utilizing shelves and racks can keep equipment off the floor and within easy reach, while a large enough work surface accommodates your equipment and any tools or accessories you might need during operation or maintenance.

Equipment Placement

The placement of your equipment can significantly impact both efficiency and comfort. For optimal operation, place your transceiver at eye level or slightly below, ensuring it is within easy reach. This setup reduces strain on your neck and eyes during long sessions. Position your power supply near the transceiver but ensure it has adequate ventilation to prevent overheating. Avoid placing it under your work surface where it might obstruct airflow or be difficult to access.

If you use a computer for logging, digital modes, or rig control, it should be within easy reach as well. Ensure the monitor is at a comfortable viewing height to prevent eye strain. If necessary, use a monitor stand to achieve the correct height.

Cable Management

Effective cable management is crucial for maintaining a clean and efficient station. Poorly managed cables can lead to signal interference, accidental disconnections, and increased wear and tear on your equipment. Label all cables to make it easier to identify connections when troubleshooting or reconfiguring your station. Use cable ties, clips, and channels to route cables neatly along walls, under desks, or through racks, and avoid running cables parallel to power lines to reduce the risk of electromagnetic interference (EMI). Use cables of appropriate lengths to minimize excess slack, coiling any extra length neatly and securing it with cable ties to prevent tangling.

Ventilation and Temperature Control

Maintaining a cool environment is essential for the longevity and performance of your equipment. High temperatures can cause electronic components to degrade faster and increase the risk of equipment failure. Ensure that your equipment is placed in a way that allows for adequate airflow, avoiding stacking components directly on top of each other. Leave space around equipment for ventilation, and consider using fans or other cooling systems to keep the air circulating in your station. In particularly hot environments, air conditioning may be necessary to maintain a comfortable and safe operating temperature. Use temperature sensors to monitor the heat levels in your station. Some power supplies and transceivers come with built-in temperature monitoring features that can alert you to overheating issues.

Practical Example of a Station Layout

A simple layout might include shelving units to keep your transceiver and power supply organized and within easy reach. A work surface beneath the shelving can accommodate your computer and any accessories. This arrangement helps keep all essential equipment accessible and the workspace organized, enhancing both efficiency and comfort.

By focusing on careful planning, effective space utilization, optimal equipment placement, cable management, and ventilation, you can create a functional and efficient operating environment. A well-organized station enhances your ham radio experience, making it easier to operate, maintain, and enjoy your equipment.

Ergonomics and Efficiency

Ergonomics is the science of designing and arranging things people use so that the people and things interact most efficiently and safely. In the context of a ham radio station, good ergonomic practices can help prevent discomfort and injury while improving your operating efficiency.

Importance of Ergonomics

Spending long hours at your ham radio station can lead to discomfort or even injury if your setup is not ergonomically designed. Poor ergonomics can cause strain on your neck, back, eyes, and wrists. By optimizing your station's layout and your operating habits, you can create a more comfortable and efficient operating environment.

Key Ergonomic Principles

1. **Proper Seating**

    - **Chair Selection**: Choose a chair that provides good lumbar support and allows you to sit comfortably for extended periods. An adjustable chair can help you find the optimal seating position.

    - **Seating Position**: Sit with your feet flat on the floor or on a footrest, and keep your knees at a 90-degree angle. Your back should be straight and supported by the chair.

2. **Optimal Equipment Placement**

- **Transceiver and Controls**: Place your transceiver and controls within easy reach to avoid unnecessary stretching. Ideally, these should be at a height where your forearms are parallel to the floor when operating.

- **Monitor Height**: If you use a computer, the monitor should be at eye level to prevent neck strain. Use a monitor stand if necessary to achieve the correct height.

3. **Desk and Work Surface**

    - **Height and Depth**: Ensure that your desk or work surface is at a comfortable height and depth. Your forearms should rest comfortably on the surface when operating.

    - **Organized Workspace**: Keep your workspace organized and free of clutter. Frequently used items should be within easy reach, while less frequently used items can be stored away.

4. **Lighting**

    - **Adequate Illumination**: Ensure that your station is well-lit to reduce eye strain. Use a combination of ambient and task lighting to achieve optimal illumination.

    - **Adjustable Lighting**: Consider using adjustable lamps that allow you to direct light where it is needed most. Avoid glare on monitors and other reflective surfaces.

Efficiency Tips

1. **Workflow Optimization**

    - **Task Grouping**: Group similar tasks together to minimize movement and maximize efficiency. For example, keep all logging and digital mode equipment in one area.

    - **Pre-Planning**: Plan your operating sessions in advance. Have all necessary equipment and materials ready to avoid interruptions and inefficiencies.

2. **Cable Management**

    - **Reduced Clutter**: Organized cables not only look better but also reduce the risk of tripping and make it easier to troubleshoot issues.

    - **Quick Access**: Label and route cables in a way that allows for quick access and easy changes. This can save time when adding new equipment or reconfiguring your setup.

3. **Ergonomic Accessories**

    - **Wrist Rests**: Use wrist rests to support your wrists while typing or using a mouse. This can help prevent repetitive strain injuries.

    - **Monitor Stands**: Use monitor stands or adjustable mounts to position your monitors at the correct height.

4. **Breaks and Movement**

    - **Regular Breaks**: Take regular breaks to stand, stretch, and move around. This helps prevent stiffness and reduces the risk of repetitive strain injuries.
    - **Exercise**: Incorporate light exercises or stretches into your routine to keep your body flexible and reduce fatigue.

Example Ergonomic Checklist

Ergonomic Element	Description	Check
Chair	Adjustable, with lumbar support	
Monitor Height	At eye level	
Transceiver Position	Within easy reach, forearms parallel	
Desk Height	Comfortable height for forearm support	
Lighting	Adequate, with adjustable task lighting	
Cable Management	Organized and labeled	
Regular Breaks	Scheduled breaks for stretching	

Safety Considerations

Safety is a paramount concern when setting up and maintaining your ham radio station. Ensuring that your station is safe can prevent accidents, protect your equipment, and safeguard your health.

Electrical Safety

1. **Proper Grounding**

    - **Grounding Equipment**: Ensure that all equipment is properly grounded to prevent electrical shocks and reduce the risk of damage from electrical surges. Use a common grounding point for all devices.
    - **Antenna Grounding**: Ground your antennas to protect against lightning strikes. Use a grounding rod installed at least eight feet into the ground and connect it with heavy-gauge wire.

2. **Safe Wiring Practices**

- **Wire Management**: Organize and secure all wires to prevent tripping hazards and accidental disconnections. Use cable ties and channels to keep wires neat and out of the way.
- **Quality Components**: Use high-quality cables and connectors that are rated for your equipment's power levels. Avoid using damaged or frayed cables.

3. **Power Supply Safety**
    - **Ventilation**: Ensure that your power supply has adequate ventilation to prevent overheating. Do not cover or obstruct the cooling vents.
    - **Circuit Protection**: Use fuses or circuit breakers to protect your equipment from electrical faults. Ensure that the power supply is properly rated for your equipment.

RF Exposure

1. **Understanding RF Exposure**
    - **Potential Hazards**: Radio frequency (RF) exposure can pose health risks if not properly managed. Long-term exposure to high levels of RF radiation can cause tissue damage and other health issues.
    - **Regulations**: Familiarize yourself with local regulations and guidelines on RF exposure limits. The Federal Communications Commission (FCC) provides guidelines for safe RF exposure levels.

2. **Minimizing RF Exposure**
    - **Antenna Placement**: Place antennas away from living areas and ensure they are at a safe distance from people. The height and distance of the antenna can significantly reduce exposure levels.
    - **Power Levels**: Use the minimum power necessary to make contact. Reducing power levels not only minimizes RF exposure but also reduces interference with other electronic devices.

3. **Monitoring RF Exposure**
    - **RF Meters**: Use RF exposure meters to measure the levels of RF radiation in your station and surrounding areas. This can help you ensure that exposure levels remain within safe limits.
    - **Regular Checks**: Periodically check RF exposure levels, especially after making changes to your station setup or antenna configuration.

Fire Safety

1. **Fire Prevention**

- **No Overloading**: Avoid overloading electrical outlets and circuits. Spread the load across multiple outlets and use surge protectors.
- **Avoid Flammable Materials**: Keep flammable materials away from your station. Ensure that your operating area is clean and free of clutter.

2. **Fire Detection**
   - **Smoke Detectors**: Install smoke detectors in your operating area to provide early warning in case of a fire. Test them regularly to ensure they are functioning properly.
   - **Fire Extinguishers**: Keep a fire extinguisher within reach of your station. Ensure that it is suitable for electrical fires and that you know how to use it.

3. **Emergency Preparedness**
   - **Evacuation Plan**: Have an evacuation plan in place in case of a fire. Ensure that all household members are aware of the plan and know the escape routes.
   - **Regular Drills**: Conduct regular fire drills to ensure that everyone knows what to do in an emergency.

Example Safety Checklist

Safety Element	Description	Check
Grounding	Proper grounding for all equipment	
Wire Management	Organized and secured wires	
Ventilation	Adequate ventilation for power supply	
Circuit Protection	Fuses or circuit breakers in place	
Antenna Placement	Safe distance from living areas	
Power Levels	Minimum necessary power used	
RF Meters	Regular RF exposure checks	
Smoke Detectors	Installed and tested	
Fire Extinguishers	Accessible and suitable for electrical fires	
Evacuation Plan	Plan in place and practiced	

Maintenance and Upkeep

Regular maintenance and upkeep of your ham radio station are essential for ensuring the longevity and performance of your equipment. A well-maintained station operates more efficiently and reliably, reducing the likelihood of unexpected failures.

Routine Maintenance Tasks

1. **Cleaning Equipment**

    - **Dust Removal**: Dust can accumulate on and inside your equipment, leading to overheating and potential damage. Use a soft, dry cloth to wipe down external surfaces regularly. For internal components, use compressed air to blow out dust.

    - **Contact Cleaning**: Over time, contacts and connectors can become dirty or corroded, leading to poor connections. Use contact cleaner to clean these areas and ensure a good connection.

2. **Visual Inspections**

    - **Cable Checks**: Regularly inspect all cables for signs of wear, damage, or fraying. Replace any damaged cables immediately to prevent signal loss or electrical hazards.

    - **Connector Inspections**: Check all connectors for tightness and signs of corrosion. Loose or corroded connectors can cause intermittent connections and signal degradation.

3. **Antenna Maintenance**

    - **Physical Inspection**: Periodically inspect your antennas for damage, corrosion, or loose elements. Ensure that all mounting hardware is secure and that the antenna is properly aligned.

    - **SWR Checks**: Regularly check the standing wave ratio (SWR) of your antennas to ensure they are properly tuned. High SWR readings can indicate problems with the antenna or feedline.

Software and Firmware Updates

1. **Keeping Software Current**

    - **Logging and Control Software**: Regularly update your logging and rig control software to benefit from the latest features and bug fixes. Check the software developer's website for updates and installation instructions.

    - **Digital Mode Software**: Ensure that your digital mode software is up to date. Updates often include new modes, improved decoding algorithms, and enhanced functionality.

2. **Firmware Updates**

    - **Transceiver Firmware**: Manufacturers periodically release firmware updates for transceivers to improve performance, add features, and fix bugs. Check your transceiver's firmware version and update it if a newer version is available.

    - **Interface Devices**: Update the firmware of any interface devices, such as sound card interfaces or antenna tuners, to ensure compatibility and optimal performance.

Preventative Maintenance

1. **Scheduled Maintenance**

    - **Routine Schedule**: Establish a routine maintenance schedule based on the manufacturer's recommendations and your operating conditions. Regular maintenance can help identify potential issues before they become major problems.

    - **Documentation**: Keep a maintenance log to track all maintenance activities, including cleaning, inspections, and software updates. This helps you stay organized and provides a history of your station's upkeep.

2. **Component Testing**

    - **Power Supply Testing**: Regularly test your power supply to ensure it is providing the correct voltage and current. Use a multimeter to check the output and look for any signs of fluctuation or instability.

    - **Battery Checks**: If your station includes battery backup, regularly check the condition and charge level of the batteries. Replace any batteries that show signs of wear or reduced capacity.

3. **Backup Systems**

    - **Data Backup**: Regularly back up your logging software and any other critical data. Use external drives or cloud storage to ensure your data is safe in case of a hardware failure.

    - **Redundant Systems**: Consider having backup equipment for critical components, such as a spare power supply or an additional transceiver. This can help minimize downtime if a component fails.

Example Maintenance Schedule

Maintenance Task	Frequency	Description
Dust Removal	Monthly	Wipe down equipment and use compressed air
Contact Cleaning	Quarterly	Clean contacts and connectors

Maintenance Task	Frequency	Description
Cable and Connector Check	Quarterly	Inspect for wear, damage, and corrosion
Antenna Inspection	Biannually	Check for physical damage and secure hardware
SWR Checks	Monthly	Measure SWR to ensure proper tuning
Software Updates	Monthly	Update logging, control, and digital mode software
Firmware Updates	Biannually	Update transceiver and interface firmware
Power Supply Testing	Quarterly	Check voltage and current output
Battery Checks	Monthly	Inspect and test battery condition
Data Backup	Weekly	Back up logging software and critical data

Routine Checks and Troubleshooting

Routine checks and troubleshooting are vital for maintaining the performance and reliability of your ham radio station. Regular checks help identify potential issues early, while effective troubleshooting techniques can quickly resolve problems and minimize downtime.

Routine Checks

1. **Performance Monitoring**

    - **Signal Reports**: Regularly ask for signal reports from other operators to gauge the quality of your transmissions. Pay attention to reports of weak signals, distortion, or interference, as these can indicate underlying issues.

    - **Audio Quality**: Monitor your audio quality during transmissions. Listen for any signs of hum, noise, or distortion, which can be caused by grounding issues, faulty cables, or equipment malfunctions.

2. **Antenna System Checks**

    - **SWR Readings**: Measure the standing wave ratio (SWR) of your antennas regularly to ensure they remain properly tuned. An increase in SWR can indicate a problem with the antenna, feedline, or connectors.

    - **Visual Inspections**: Perform visual inspections of your antennas and feedlines to check for physical damage, corrosion, or loose connections.

3. **Power Supply Checks**

    - **Voltage and Current**: Use a multimeter to check the voltage and current output of your power supply. Ensure that it remains stable and within the specified range for your equipment.

    - **Temperature**: Monitor the temperature of your power supply and other equipment. Overheating can be a sign of inadequate ventilation or a failing component.

Troubleshooting Techniques

1. **Systematic Approach**

    - **Identify the Problem**: Clearly define the issue you are experiencing. Gather as much information as possible, including symptoms, error messages, and recent changes to your station.

    - **Isolate the Issue**: Use a systematic approach to isolate the problem. Break down your station into its individual components and test each one independently to identify the source of the issue.

2. **Common Issues and Solutions**

    - **No Power**: If your equipment does not power on, check the power supply connections, fuses, and circuit breakers. Ensure that the power supply is functioning correctly and providing the necessary voltage.

    - **Poor Signal Quality**: If you are receiving poor signal reports, check your antenna system for high SWR, loose connections, or damaged feedlines. Ensure that your transceiver settings, such as power output and modulation, are correct.

    - **Interference**: If you experience interference, try to identify the source. Common culprits include nearby electronic devices, power lines, and other radio transmitters. Use filters and proper grounding techniques to mitigate interference.

3. **Documentation and Resources**

    - **Manuals and Guides**: Keep the manuals and guides for your equipment readily available. They often include troubleshooting sections that can provide valuable insights and solutions.

    - **Online Forums and Communities**: Participate in online forums and ham radio communities. These platforms are excellent resources for troubleshooting advice and support from experienced operators.

Practical Example: Troubleshooting High SWR

Suppose you notice that your SWR readings have suddenly increased. Here's a step-by-step approach to troubleshooting the issue:

1. **Check Connections**
   - **Inspect Connectors**: Check all connectors along the feedline for tightness and signs of corrosion. Loose or corroded connectors can cause high SWR.
   - **Cable Continuity**: Use a multimeter to test the continuity of your feedline. A break or short in the cable can lead to high SWR readings.

2. **Inspect the Antenna**
   - **Physical Damage**: Visually inspect the antenna for any signs of damage, such as bent elements or broken insulators. Ensure that all elements are properly aligned and secure.
   - **Environmental Factors**: Consider any recent environmental changes, such as storms or new construction, that could have affected your antenna's performance.

3. **Test with a Dummy Load**
   - **Dummy Load Test**: Connect a dummy load to your transceiver and measure the SWR. A dummy load provides a perfect impedance match and should show a very low SWR. If the SWR is low with the dummy load, the problem likely lies with the antenna system.

4. **Systematic Replacement**
   - **Replace Components**: Systematically replace components in the antenna system, starting with the feedline, to identify the faulty part. Testing each component individually can help isolate the issue.

Upgrading Equipment

Upgrading your ham radio equipment can be a thrilling and rewarding experience, offering new capabilities, improved performance, and enhanced enjoyment of the hobby. Whether you're looking to expand your frequency coverage, add new features, or simply modernize your setup, a thoughtful approach to upgrading can make a significant difference.

Reasons for Upgrading

One of the primary reasons to upgrade your equipment is to enhance performance. As technology advances, newer transceivers offer better sensitivity, improved filtering, and more reliable signal quality. For instance, upgrading to a transceiver with advanced digital signal processing (DSP) can drastically reduce noise and interference, allowing you to hear weaker signals more clearly. This can be particularly beneficial for DXing and contesting, where every bit of signal clarity counts.

Adding new features is another compelling reason to consider an upgrade. Modern transceivers often come with built-in capabilities for digital modes such as FT8, PSK31, and RTTY. These modes have become increasingly popular due to their efficiency and ability to make contacts under challenging conditions. Additionally, features like automatic antenna tuning, built-in SWR meters, and remote operation capabilities can simplify your operating experience and open up new possibilities for your station.

Expanding your frequency coverage is also a significant motivation for upgrading. If your current transceiver only covers a limited range of frequencies, upgrading to a multi-band transceiver can give you access to more bands and modes. This expansion allows you to explore different parts of the radio spectrum, make contacts on new bands, and take advantage of varying propagation conditions.

Planning Your Upgrade

Before diving into an upgrade, it's essential to assess your needs and goals. Start by identifying what you hope to achieve with the new equipment. Are you looking to improve your signal quality, add digital mode capabilities, or increase your operating range? Understanding your objectives will help you make informed decisions about which equipment to purchase.

Budget is another crucial consideration. Determine how much you're willing to spend and prioritize your needs accordingly. It's important to factor in not only the cost of the new equipment but also any additional accessories, installation costs, and potential upgrades to your station layout to accommodate the new gear.

Research is key to a successful upgrade. Spend time reading reviews, seeking recommendations from fellow ham radio operators, and comparing the specifications of different models. Look for equipment that has received positive feedback for reliability, ease of use, and performance. Online forums, ham radio clubs, and product reviews are excellent resources for gathering information and getting advice from experienced operators.

Implementing the Upgrade

Once you've selected your new equipment, careful planning and execution are essential for a smooth transition. Start by backing up your current settings and data, such as logging software and configuration files, to avoid losing important information during the upgrade process.

When the new equipment arrives, follow the manufacturer's instructions for installation and setup. Pay close attention to any specific configuration steps or calibration procedures. For example, if you're upgrading to a transceiver with digital mode capabilities, you'll need to install and configure digital mode software on your computer, such as WSJT-X or FLDIGI.

Testing is a critical part of the upgrade process. Before integrating the new equipment into your full setup, test each component individually to ensure it's functioning correctly. This includes checking power levels, frequency accuracy, and signal quality. Make any necessary adjustments to fine-tune the settings and optimize performance.

CHAPTER 10 – Operating Awards and Contests

Introduction to Awards and Contests

Participating in awards and contests is a highly rewarding aspect of ham radio. Not only do these activities provide a sense of achievement, but they also enhance your operating skills, expand your knowledge, and increase your enjoyment of the hobby. Whether you are a seasoned operator or new to ham radio, there is a wide range of awards and contests to explore.

Awards in ham radio are typically earned by achieving specific milestones, such as making a certain number of contacts with different countries or states. These awards recognize the operator's dedication, skill, and perseverance. They can range from relatively easy to highly challenging, providing goals for operators at all levels.

One of the most prestigious awards is the DX Century Club (DXCC), which is awarded for making confirmed contacts with at least 100 different countries. This award encourages operators to develop their DXing (long-distance communication) skills and to learn about propagation, antennas, and operating techniques. Other popular awards include Worked All States (WAS), which requires contacts with all 50 U.S. states, and Worked All Continents (WAC), which involves contacting each of the six inhabited continents.

Contests, on the other hand, are competitive events where operators aim to make as many contacts as possible within a set period. These events can range from small, local contests to large, international competitions involving thousands of operators. Contests test your operating skills, equipment, and strategies under varying conditions, often pushing you to improve your station and techniques.

Field Day is one of the most well-known contests, especially in North America. It is an annual event organized by the American Radio Relay League (ARRL) and serves both as a contest and an emergency preparedness exercise. Operators set up portable stations and aim to make as many contacts as possible within a 24-hour period. Field Day is a fantastic opportunity to practice setting up and operating in field conditions, learn from other operators, and enjoy the camaraderie of the ham radio community.

Another major contest is the ARRL Sweepstakes, which focuses on making contacts within the United States and Canada. The goal is to work as many stations as possible on different bands, using a unique exchange that includes information such as the operator's call sign, serial number, and location. This contest emphasizes both speed and accuracy, challenging operators to manage high rates of contacts while minimizing errors.

To succeed in awards and contests, preparation is key. Start by researching the specific requirements and rules for the awards or contests you are interested in. For awards, ensure you understand the verification process, which typically involves submitting QSL cards or using an online confirmation system like Logbook of The World (LoTW). For contests, study the rules, scoring systems, and any special conditions that might apply.

It's also important to have a well-equipped and reliable station. While you don't need the latest and greatest equipment to participate, having a solid transceiver, good antennas, and effective logging software will make a significant difference. Practice operating your station efficiently, and become familiar with the features and settings of your equipment.

Additionally, develop good operating habits, such as listening carefully before transmitting, being courteous to other operators, and maintaining accurate logs. These practices not only improve your chances of success but also enhance the overall experience for everyone involved.

Popular Awards: DXCC, WAS, WAC

Earning awards in ham radio is a fulfilling way to measure your progress and achievements. Among the most sought-after awards are the DX Century Club (DXCC), Worked All States (WAS), and Worked All Continents (WAC). Each of these awards presents its unique challenges and requires different levels of dedication and skill.

DX Century Club (DXCC)

The DX Century Club, or DXCC, is one of the most prestigious awards in ham radio. To earn this award, you must make confirmed contacts with at least 100 different entities recognized by the ARRL. These entities are generally countries, but can also include certain territories and regions. The DXCC program encourages operators to improve their DXing skills, learn about propagation, and optimize their station for long-distance communication.

- **Challenges**: Achieving DXCC requires patience and perseverance. Propagation conditions, time zones, and political boundaries can make contacting certain entities difficult. Additionally, confirming these contacts often involves exchanging QSL cards or using electronic confirmation services like Logbook of The World (LoTW).

- **Levels**: After reaching the initial 100 entities, operators can continue to earn endorsements for additional entities, bands, and modes. This makes DXCC a lifelong pursuit for many hams.

Worked All States (WAS)

The Worked All States award, or WAS, is another highly regarded award, particularly among U.S. operators. To earn this award, you must make confirmed contacts with each of the 50 states in the U.S. WAS is an excellent way to practice operating on different bands and modes, as conditions vary greatly across the country.

- **Challenges**: Some states have fewer active operators, making them harder to contact. Additionally, propagation conditions can affect the ability to reach certain areas, requiring strategic planning and sometimes waiting for the right conditions.

- **Modes and Bands**: Like DXCC, WAS offers endorsements for different bands and modes, allowing operators to earn multiple WAS awards by working all states on different frequencies or using different transmission methods.

Worked All Continents (WAC)

The Worked All Continents award, or WAC, requires confirmed contacts with each of the six inhabited continents: Africa, Asia, Europe, North America, Oceania, and South America. WAC is a great introduction to DXing, as it involves making long-distance contacts but is generally easier to achieve than DXCC.

- **Challenges**: Propagation and time zones play significant roles in achieving WAC. Understanding the best times and frequencies to contact different continents is crucial.

- **Endorsements**: WAC can also be earned on different bands and modes, providing additional challenges for experienced operators.

Each of these awards—DXCC, WAS, and WAC—offers a rewarding goal for ham radio operators. They not only recognize your achievements but also encourage you to continuously improve your skills and expand your horizons in the world of amateur radio.

Major Contests: Field Day, Sweepstakes

Contests are an exciting aspect of ham radio, providing operators with the opportunity to test their skills, equipment, and strategies under competitive conditions. Two of the most prominent contests in the amateur radio world are Field Day and the ARRL Sweepstakes. These events draw thousands of participants each year and offer unique challenges and rewards.

Field Day

Field Day is arguably the most well-known contest in North America, organized annually by the American Radio Relay League (ARRL). It takes place over a 24-hour period on the fourth full weekend in June. Field Day serves as both a contest and an emergency preparedness exercise, encouraging operators to set up portable stations and operate under simulated emergency conditions.

- **Objectives**: The primary goal of Field Day is to make as many contacts as possible within the 24-hour period. However, it also emphasizes the ability to set up and operate equipment in the field, using alternative power sources such as batteries, generators, or solar panels.

- **Categories**: Participants can operate in various categories based on their setup, including single operator, multi-operator, and club stations. There are also classes based on power levels and the type of location (e.g., home, outdoor).

- **Community Aspect**: Field Day is a highly social event, often involving local clubs and groups. It provides an excellent opportunity for learning, sharing knowledge, and building camaraderie among operators.

ARRL Sweepstakes

The ARRL Sweepstakes is another major contest, renowned for its challenging format and unique exchange requirements. It is held twice a year, with separate CW (Morse code) and phone (voice) events in November. The contest focuses on making contacts within the United States and Canada.

- **Objectives**: The goal of Sweepstakes is to work as many stations as possible on different bands. Each contact must include a detailed exchange, which consists of a serial number, precedence (operating category), call sign, check (the last two digits of the operator's first year licensed), and the operator's section (geographic area).

- **Complexity**: The complexity of the exchange requires operators to be accurate and efficient, testing both their operating skills and their ability to manage high rates of contacts.

- **Strategies**: Successful Sweepstakes participants often employ various strategies, such as band planning, managing pile-ups, and optimizing their station setup for rapid exchanges.

Benefits of Participation

Field Day and the ARRL Sweepstakes bring a multitude of benefits to those who take part. These events create a well-organized setting that allows operators to refine their skills, whether it's mastering the art of managing pile-ups, keeping precise logs, or enhancing the performance of their equipment. Moreover, these contests motivate participants to try out various bands and modes, broadening their understanding and proficiency in amateur radio.

Engaging in major contests like Field Day and the ARRL Sweepstakes is a highly gratifying activity that can greatly enrich your ham radio journey. These events not only put your technical and operational skills to the test but also help build a strong sense of community and teamwork among radio enthusiasts. Whether your goal is to achieve a top score or simply to revel in the excitement of the competition, both Field Day and Sweepstakes offer valuable experiences for every ham radio operator.

Strategies for Success

Success in ham radio contests and awards requires more than just having good equipment; it demands effective strategies and meticulous preparation. Whether you're aiming for high scores in contests or prestigious awards, having a well-thought-out plan can significantly enhance your chances of success.

The foundation of any successful contest or award pursuit is thorough preparation. This begins with familiarizing yourself with the specific rules and requirements. Knowing the details, such as the scoring system, exchange requirements, and operating times, is crucial. Spend time researching past contests or awards to understand trends and strategies that have worked for others. Insights from online forums, ham radio clubs, and publications can be invaluable.

Equally important is ensuring that your station is well-prepared and optimized. This means checking all your equipment, updating your software, and conducting test runs to identify and resolve any potential issues. Pay particular attention to your antenna setup, as it plays a crucial role in your ability to make contacts.

Once the contest or award pursuit begins, efficient operating techniques are key. Developing a band plan can be very helpful. Propagation conditions vary throughout the day, so having a plan that takes advantage of peak times for each band can improve your contact rate. It's also important to stay flexible and ready to adjust your plan based on real-time conditions. Maintaining accurate logs is essential, and using logging software can streamline the process and reduce the risk of errors. Double-checking entries ensures all information is correct, as mistakes can lead to disqualifications or missed points.

Monitoring your contact rate and adjusting your strategy to maintain a high rate of contacts is also crucial. If a particular band becomes less productive, be prepared to switch to another band or mode to keep your rate up. It's often more efficient to move on from difficult contacts and return later.

In many contests, maximizing points and multipliers is critical. Focusing on working multipliers, such as different states, countries, or zones, early in the contest can have a significant impact on your score. Prioritize working new multipliers whenever possible, even if it means shifting bands or modes. Additionally, identifying and prioritizing high-value contacts that offer more points can boost your score significantly. For example, some contests award more points for long-distance (DX) contacts or contacts on higher frequency bands.

Staying organized and focused throughout the contest or award pursuit is vital. Using checklists can ensure you don't forget important tasks or details, such as pre-contest setup, band changes, and post-contest log submission. Regular breaks to rest and recharge are essential. Staying hydrated and having snacks on hand can help maintain your energy levels and focus. It's important to take care of your physical and mental well-being during long operating sessions.

After the contest, take the time to review your performance. Analyzing what worked well and identifying areas for improvement is crucial. Reviewing your log and noting any mistakes or missed opportunities can help refine your strategies for future events.

Networking with other operators and learning from their experiences can also provide valuable insights and motivation. Joining ham radio clubs or contesting teams allows you to share knowledge, resources, and support. Working with a team can enhance your learning experience and provide opportunities for collaboration. Seeking mentorship from more experienced operators can accelerate your progress and help you avoid common pitfalls.

Contesting Techniques

Contesting is one of the most exhilarating aspects of ham radio, offering operators the chance to test their skills, equipment, and strategies against others. Successful contesting requires a blend of technical proficiency, strategic planning, and quick decision-making. Here are some techniques that can help you excel in contests.

The first step to success in contesting is to thoroughly understand the contest rules. Each contest has its own set of rules regarding the exchange, scoring, multipliers, and operating times. Familiarizing yourself with these rules helps avoid mistakes that can lead to disqualification or lost points. It's also important to plan your operating times strategically, taking into account the propagation conditions and peak activity periods for different bands.

One of the most critical skills in contesting is managing pile-ups. A pile-up occurs when multiple stations are calling you at the same time, and handling them efficiently is key to maintaining a high contact rate. Listening carefully and quickly picking out call signs from the noise is essential. A good technique is to call for specific parts of the pile-up, such as by asking for stations from a particular region or with a certain suffix in their call sign. This helps to manage the chaos and make the process more orderly.

Rate management is another crucial aspect of contesting. Your rate is the number of contacts you make per hour, and keeping this number high is essential for a competitive score. Monitoring your rate and making adjustments as needed can significantly impact your performance. For example, if your current band is becoming less productive, be ready to switch to another band or mode to maintain your rate. Using two radios (SO2R, or Single Operator Two Radios) can also help by allowing you to search for new contacts on one band while transmitting on another.

Efficient logging is vital for contest success. Using software specifically designed for contest logging can help you keep track of your contacts, multipliers, and score in real time. These programs often have features that automate many aspects of the logging process, reducing the chances of errors and saving you valuable time. Double-checking entries and ensuring accuracy in your logs is crucial, as mistakes can lead to penalties or disqualification.

Operating ergonomics play a significant role during long contest sessions. Ensuring that your operating position is comfortable and that your equipment is easily accessible can help you maintain efficiency and reduce fatigue. Having snacks and drinks on hand, taking regular short breaks, and stretching can keep your energy levels up and your mind sharp.

Understanding and leveraging propagation is another key technique. Knowing when and where different bands are open can help you plan your band changes and maximize your contact rate. Tools like propagation prediction software and real-time monitoring of beacons and DX clusters can provide valuable information on current conditions.

Networking with other contesters and learning from their experiences can also enhance your skills. Participating in online forums, joining contesting clubs, and reading articles and books on contesting techniques can provide valuable insights and tips. Observing seasoned operators, either through online streams or in-person, can also teach you effective methods and strategies.

Award Hunting Tips

Pursuing ham radio awards can be an exciting and fulfilling endeavor. Awards such as DXCC, WAS, and WAC not only provide a sense of accomplishment but also encourage you to expand your skills and knowledge. Here are some tips to help you effectively hunt for and achieve these awards.

Setting Clear Goals

The first step in award hunting is setting clear, achievable goals. Decide which awards you want to pursue and familiarize yourself with their specific requirements. Understanding the criteria for each award helps you create a focused plan. For example, if you're aiming for the DX Century Club (DXCC), you need to make confirmed contacts with at least 100 different entities. Knowing this, you can prioritize contacting new entities during your operating sessions.

Utilizing Effective Tools

Effective award hunting often involves using the right tools and resources. Logging software is essential for tracking your contacts and ensuring you meet the award requirements. Programs like Ham Radio Deluxe or N1MM Logger+ can help you keep detailed records and generate reports on your progress. Additionally, using online services like Logbook of The World (LoTW) simplifies the confirmation process, allowing you to verify contacts electronically.

Propagation prediction tools are also invaluable. Understanding propagation conditions helps you know when and where to operate to reach specific areas. Tools like VOACAP provide detailed predictions based on current solar conditions, helping you plan your operating schedule for optimal results.

Joining Clubs and Networks

Joining ham radio clubs and networks can greatly enhance your award hunting efforts. Clubs often have resources, events, and experienced members who can offer advice and support. Participating in club activities and nets increases your chances of making the necessary contacts for your awards. Additionally, being part of a club can provide opportunities to learn from more experienced operators and share in the camaraderie of the hobby.

Operating Strategies

Adopting effective operating strategies is key to successful award hunting. One important strategy is to be active during major contests and DXpeditions. These events draw operators from around the world, increasing your chances of making contacts with new entities. Even if you are not competing in the contest, participating as a casual operator can help you work towards your award goals.

Another strategy is to vary your operating times and bands. Different bands and times of day offer varying propagation conditions, allowing you to reach different parts of the world. For example, higher HF bands like 20 meters and 15 meters are often open during the day, while lower bands like 40 meters and 80 meters can be more effective at night.

Being patient and persistent is crucial. Some entities or states may be rare or difficult to contact, requiring multiple attempts. Keep trying, and don't get discouraged by initial setbacks. Utilizing QSL card services and online databases to track active stations from rare entities can also improve your chances.

Record-Keeping and Verification

Maintaining accurate and detailed records is essential for award hunting. Ensure that your logs are up to date and that you have confirmed each contact either through QSL cards or electronic confirmation. Regularly review your logs to identify any gaps or errors.

For awards like DXCC, WAS, and WAC, using services like LoTW can simplify the verification process. Ensure that your log entries are consistent and that all necessary information, such as call signs, dates, and frequencies, is correctly recorded.

Leveraging Technology

Modern technology can greatly enhance your award hunting efforts. Using software-defined radios (SDRs) and digital modes can open up new possibilities for making contacts. Digital modes like FT8 and JT65 are particularly effective for working weak signals and achieving long-distance contacts under challenging conditions.

Staying Informed

Keeping up with the latest news and developments in the ham radio community is important. Subscribe to ham radio magazines, join online forums, and follow social media groups to stay informed about upcoming events, new technologies, and tips from other operators. This knowledge can provide valuable insights and opportunities to advance your award hunting goals.

CHAPTER 11 – Advanced Topics

Homebrewing Equipment

Homebrewing equipment is a fascinating and rewarding aspect of amateur radio, allowing operators to build their own gear and tailor their stations to their specific needs and preferences. This section delves into the world of homebrewing, exploring the benefits, challenges, and essential steps involved in creating your own radio equipment.

Homebrewing refers to the practice of designing and constructing your own radio equipment from scratch or from kits. This can include anything from simple antennas to complex transceivers. The primary appeal of homebrewing is the ability to customize and experiment with your equipment, gaining a deeper understanding of radio principles and enhancing your technical skills.

Benefits of Homebrewing

One of the main benefits of homebrewing is the sense of accomplishment that comes from using equipment you built yourself. This hands-on experience provides invaluable insights into the workings of radio gear, helping you to become a more knowledgeable and skilled operator. Additionally, homebrewing can be cost-effective, as building your own equipment often costs less than purchasing commercial gear. It also allows for greater flexibility and innovation, as you can modify and adapt your designs to meet your specific needs.

Getting Started

Before embarking on a homebrewing project, it is important to gather the necessary tools and materials. Basic tools include a soldering iron, multimeter, wire cutters, pliers, and screwdrivers. Depending on the complexity of the project, you may also need specialized tools such as an oscilloscope, signal generator, or spectrum analyzer.

Start with a simple project, such as building a basic antenna or a crystal radio. These projects provide a solid foundation in electronics and radio principles, allowing you to develop the skills needed for more complex projects. As you gain confidence and experience, you can move on to more advanced projects such as building a transceiver or amplifier.

Schematic Diagrams and Circuit Boards

Understanding schematic diagrams is essential for homebrewing. Schematics provide a visual representation of the electrical connections and components in a circuit, guiding you through the construction process. Take the time to familiarize yourself with common symbols and conventions used in schematics, as this will make it easier to follow and modify circuit designs.

Creating or obtaining printed circuit boards (PCBs) is another important step. PCBs provide a solid foundation for mounting and connecting components, ensuring a reliable and organized build. You can design your own PCBs using software tools or purchase pre-designed boards for specific projects.

Challenges and Solutions

Homebrewing can be challenging, especially for beginners. One common challenge is sourcing the right components. Many homebrewers turn to online suppliers, electronics stores, or hamfests to find the parts they need. Ensure that you purchase quality components to avoid issues with reliability and performance.

Another challenge is troubleshooting. Even with careful planning and execution, problems can arise. Developing good troubleshooting skills is essential. Start by checking for common issues such as loose connections, incorrect component placement, or solder bridges. Using diagnostic tools like a multimeter or oscilloscope can help you pinpoint and resolve issues.

Community and Resources

The ham radio community is a valuable resource for homebrewers. Joining a local radio club or online forum can provide support, advice, and inspiration. Many clubs offer workshops or mentoring programs for beginners, helping you to learn new skills and techniques. Online resources such as instructional videos, blogs, and project guides can also be immensely helpful.

Example Table: Basic Tools for Homebrewing

Tool	Description
Soldering Iron	For soldering components to circuit boards
Multimeter	For measuring voltage, current, and resistance
Wire Cutters	For cutting wires and component leads
Pliers	For gripping and bending wires

Tool	Description
Screwdrivers	For assembling and disassembling enclosures
Oscilloscope	For visualizing electronic signals
Signal Generator	For generating test signals
Spectrum Analyzer	For analyzing signal frequencies and strengths

Building Your Own Gear

Building your own radio gear is one of the most rewarding aspects of amateur radio. This process not only enhances your understanding of electronics and radio principles but also allows you to customize equipment to your specific needs and preferences. Whether you are constructing a simple antenna or a complex transceiver, the skills and knowledge gained through homebrewing are invaluable.

Starting with Simple Projects

For beginners, starting with simple projects is a wise approach. Building basic equipment like antennas, dummy loads, or simple transmitters can provide a solid foundation in electronics and radio theory. For example, constructing a dipole antenna is a straightforward project that teaches essential skills such as soldering, measuring, and understanding resonance.

A crystal radio is another excellent starter project. This type of radio receiver uses the power of radio signals to produce sound without the need for external power sources. Building a crystal radio introduces you to basic concepts like tuning circuits, diodes, and the importance of antenna and ground connections.

Understanding Schematics and Components

As you progress to more complex projects, understanding schematic diagrams becomes crucial. Schematics are the blueprints of electronic circuits, showing how components are connected. Familiarize yourself with common symbols and learn to read and interpret these diagrams. This skill is fundamental for successful homebrewing.

Components such as resistors, capacitors, transistors, and inductors form the building blocks of electronic circuits. Understanding the function of each component and how they interact within a circuit is key. Using a multimeter to test components and verify connections can help prevent mistakes and ensure your projects work as intended.

Assembling Your Project

Once you have a schematic and all necessary components, the next step is assembly. This process typically involves soldering components onto a printed circuit board (PCB) or constructing circuits on a breadboard for prototyping. Soldering is a critical skill in homebrewing, requiring practice to master. Ensure your solder joints are clean and secure to prevent poor connections and signal loss.

If you are designing your own circuits, consider using PCB design software. These tools allow you to create precise layouts for your circuits, which can then be manufactured or etched at home. For those not ready to design their own PCBs, many kits come with pre-made boards that simplify the assembly process.

Troubleshooting and Testing

Troubleshooting is an inevitable part of building your own gear. Even with careful planning and execution, issues can arise. Start by visually inspecting your work for obvious errors such as solder bridges, missing components, or incorrect orientations. Use a multimeter to check for continuity and proper voltages throughout the circuit.

Testing your project in stages can help identify issues early. For example, if you are building a transceiver, test the power supply, audio circuits, and RF sections separately before integrating them. This step-by-step approach simplifies troubleshooting and ensures each part of the circuit functions correctly.

Documentation and Sharing

Documenting your projects is important for future reference and for sharing with the ham radio community. Keep detailed notes on your designs, components used, and any modifications or troubleshooting steps taken. Photos and diagrams can also be helpful.

Sharing your projects online or with local ham radio clubs can provide valuable feedback and inspire others. Platforms like blogs, forums, and social media allow you to connect with other homebrewers, exchange ideas, and learn from their experiences.

Example: Building a QRP Transceiver

Building a QRP (low-power) transceiver is a popular project for more advanced homebrewers. These transceivers are designed for low power operation, typically under 5 watts, and are ideal for portable or minimalist setups.

1. **Design and Components**: Choose a proven design from reputable sources or develop your own. Gather components including a VFO (variable frequency oscillator), mixer, filter circuits, and final amplifier stage.

2. **Assembly**: Carefully assemble each stage of the transceiver, starting with the power supply and progressing through the audio and RF sections. Use shielded enclosures to minimize interference.

3. **Testing**: Test each stage independently before integrating them. Use a dummy load and signal generator to verify proper operation. Once integrated, make on-air tests to evaluate performance.

Building your own gear, whether simple or complex, is a deeply satisfying aspect of amateur radio. It enhances your technical skills, allows for customization, and fosters a deeper connection to the hobby. By starting with simple projects, understanding schematics, mastering assembly and troubleshooting techniques, and documenting your work, you can successfully create your own radio equipment and enjoy the myriad benefits of homebrewing.

Kits and Resources

Kits and resources play a crucial role in homebrewing, offering pre-packaged components and detailed instructions that simplify the construction of radio equipment. They provide a practical entry point for beginners and a convenient solution for experienced operators looking to build specific projects.

Advantages of Using Kits

Using kits for homebrewing offers several advantages. Kits come with all the necessary components, reducing the time and effort required to source parts individually. This is particularly beneficial for beginners who might find it challenging to identify and procure specific components. Additionally, kits include detailed instructions, which guide you through the assembly process step-by-step, ensuring a higher success rate.

Kits also offer an opportunity to learn by doing. As you assemble the kit, you gain hands-on experience with electronic components and circuit construction. This practical experience is invaluable in developing your understanding of electronics and improving your technical skills.

Popular Types of Kits

There are various types of kits available, catering to different interests and skill levels. Here are a few popular categories:

1. **Transceiver Kits**: These kits range from simple QRP (low-power) transceivers to more complex multi-band models. They are popular among operators who enjoy portable or minimalist setups.

2. **Antenna Kits**: Building your own antennas can be both cost-effective and educational. Antenna kits often include all necessary materials and detailed instructions for constructing effective antennas for different bands.

3. **Accessory Kits**: These kits include items such as antenna tuners, audio filters, and power supplies. They provide useful enhancements to your station setup and are generally easier to assemble than full transceivers.

Finding Quality Kits

When selecting a kit, it's important to choose one from a reputable supplier. Quality kits come with reliable components, clear instructions, and good customer support. Several well-known companies and organizations specialize in amateur radio kits, offering a wide range of products to suit different needs.

- **Elecraft**: Known for their high-quality transceiver kits, Elecraft offers a range of products from simple QRP kits to sophisticated multi-band transceivers. Their kits are designed with both performance and ease of assembly in mind.

- **Heathkit**: A classic name in the world of homebrewing, Heathkit has been providing kits for decades. They offer a variety of products, including transceivers, test equipment, and accessories.

- **QRPKits**: Specializing in low-power (QRP) kits, QRPKits provides a selection of transceivers, tuners, and accessories that are ideal for portable and minimalist operations.

Resources for Homebrewing

In addition to kits, there are numerous resources available to support your homebrewing efforts. These resources provide valuable information, guidance, and community support.

Books and Manuals: Many excellent books and manuals are available that cover various aspects of homebrewing. Titles such as "The ARRL Handbook for Radio Communications" and "The Radio Amateur's Handbook" are comprehensive resources that include detailed information on electronics, circuit design, and construction techniques.

Online Communities: Joining online forums and communities can provide a wealth of information and support. Websites like QRZ.com, eHam.net, and the QRP-L mailing list offer discussion groups where you can ask questions, share experiences, and learn from other homebrewers.

YouTube and Blogs: Video tutorials and blog posts can be extremely helpful, especially for visual learners. Many experienced homebrewers share their projects and techniques through YouTube channels and personal blogs, providing step-by-step guides and tips.

Local Clubs and Workshops: Participating in local amateur radio clubs and attending workshops can provide hands-on learning opportunities. Clubs often organize build nights or project workshops where members can work on kits together, share tools and expertise, and troubleshoot any issues that arise.

In summary, using kits and taking advantage of available resources can greatly enhance your homebrewing experience. Kits simplify the process of building your own gear by providing all necessary components and clear instructions. Meanwhile, books, online communities, video tutorials, and local clubs offer valuable support and learning opportunities. By leveraging these resources, you can successfully build a wide range of radio equipment and deepen your understanding of the hobby.

Software-Defined Radio (SDR)

Software-Defined Radio (SDR) represents a significant advancement in the field of amateur radio, transforming how operators interact with radio signals. Unlike traditional radios, which rely on hardware components for signal processing, SDR uses software to perform these tasks, offering unprecedented flexibility and functionality.

What is SDR?

SDR is a type of radio communication system where components that have typically been implemented in hardware (such as mixers, filters, amplifiers, modulators/demodulators, detectors, etc.) are instead implemented by means of software on a personal computer or embedded system. The flexibility of SDR allows it to support a wide range of frequencies and modes with minimal hardware changes.

Benefits of SDR

One of the primary benefits of SDR is its versatility. By simply updating or changing the software, you can use the same hardware to work across different bands and modes, from HF to VHF and UHF, and from AM and FM to digital modes like PSK31, FT8, and even complex spread spectrum techniques. This makes SDR an attractive option for operators who want to explore multiple facets of the hobby without investing in multiple radios.

Another significant advantage is the ability to implement advanced signal processing techniques. SDR can perform tasks such as digital filtering, noise reduction, and signal decoding with greater precision and efficiency than traditional analog methods. This results in clearer signals and better overall performance.

Popular SDR Platforms

Several popular SDR platforms are available, catering to different needs and budgets:

1. **RTL-SDR**: An affordable entry-level SDR, originally designed as a TV tuner, but repurposed for radio applications. It is popular among hobbyists for its low cost and wide frequency coverage (approximately 24 MHz to 1.7 GHz).

2. **HackRF One**: A more advanced SDR that supports a broader frequency range (1 MHz to 6 GHz) and includes features suitable for more serious experimentation and development.

3. **FlexRadio Systems**: High-end SDR transceivers like the Flex 6000 series offer top-tier performance with features designed for serious amateur radio operators, including superior dynamic range, customizable filters, and remote operation capabilities.

Setting Up an SDR Station

Setting up an SDR station involves both hardware and software components. The hardware typically includes the SDR device itself, an antenna suitable for the desired frequency range, and a computer for running the SDR software. Additional accessories like preamplifiers, filters, and upconverters might also be used to enhance performance.

The software component is where the real power of SDR comes into play. There are several popular SDR software programs, such as:

- **SDR# (SDRSharp)**: A free, Windows-based application that is user-friendly and supports a wide range of SDR hardware.

- **GNU Radio**: An open-source software toolkit that provides signal processing blocks to implement software radios. It is more complex but highly flexible and powerful.

- **HDSDR**: Another popular Windows application that offers extensive features and supports various SDR hardware.

Software	Operating System	Key Features
SDR#	Windows	User-friendly, wide hardware support
GNU Radio	Windows, Linux	Highly flexible, extensive signal processing blocks
HDSDR	Windows	Advanced features, customizable interface

Advanced Data Modes

Advanced data modes have significantly expanded the capabilities of amateur radio, allowing for more efficient communication, enhanced signal quality, and the ability to operate under challenging conditions. These modes leverage digital technology to transmit and receive data, making them an essential part of the modern ham radio landscape.

Understanding Advanced Data Modes

Advanced data modes go beyond traditional digital modes like RTTY and PSK31, offering greater efficiency, robustness, and versatility. Modes such as FT8, JT65, and WSPR have become increasingly popular among ham radio operators due to their ability to facilitate reliable communication over long distances and in weak signal conditions.

FT8: FT8, short for "Franke-Taylor design, 8-Frequency Shift Keying," is a highly efficient mode designed for making rapid, reliable contacts with minimal power and bandwidth. It operates with 15-second transmission intervals, allowing for quick exchanges even under poor propagation conditions. FT8 has revolutionized the way amateur radio operators engage in DXing, making it possible to achieve contacts that were previously unattainable.

JT65: Developed for weak signal communication, JT65 is particularly effective for moonbounce (EME) and meteor scatter. It uses long transmission intervals and strong error-correcting codes to ensure that signals can be decoded even when they are barely audible. This mode has opened up new possibilities for operators interested in experimenting with extreme long-distance communication.

WSPR: WSPR, or "Weak Signal Propagation Reporter," is a mode used for testing propagation paths and understanding radio wave behavior. It transmits low-power signals that can be received globally, allowing operators to study how signals propagate under different conditions. WSPR is an invaluable tool for those interested in the technical aspects of radio wave propagation and for optimizing their antennas and operating strategies.

Practical Applications

Advanced data modes have numerous practical applications that enhance the versatility of amateur radio operations. One significant application is in emergency communication. These modes are highly efficient and can transmit vital information with minimal power, making them ideal for emergency situations where power resources may be limited, and conditions may be adverse.

For DXing enthusiasts, advanced data modes offer a means to achieve contacts with distant stations that would be difficult or impossible using traditional modes. The efficiency of FT8, for example, allows operators to make reliable contacts across the globe with low power, making it a favorite for DXers seeking to expand their contact lists.

In the realm of experimental communication, advanced data modes provide a platform for innovation and development. Operators can experiment with new coding schemes, modulation techniques, and error correction algorithms, contributing to the advancement of digital communication technologies. This experimental aspect of amateur radio fosters a deeper understanding of radio science and inspires technological innovation.

Table: Comparison of Advanced Data Modes

Mode	Typical Use Case	Transmission Interval	Key Feature
FT8	Rapid contacts	15 seconds	High efficiency and quick exchanges
JT65	Weak signal communication	60 seconds	Strong error correction
WSPR	Propagation studies	Varies	Low power, global reach

Integrating Advanced Data Modes into Your Station

Integrating advanced data modes into your ham radio station involves both hardware and software considerations. A reliable transceiver capable of digital mode operation is essential. Additionally, a computer with the necessary software, such as WSJT-X for FT8 and JT65, or WSPR for propagation studies, is required.

Setting up the software involves configuring your computer to interface with your transceiver, ensuring that the audio levels are correctly adjusted for optimal signal transmission and reception. Many digital mode software applications provide detailed guides and support forums to assist with the setup process, making it accessible even for those new to digital modes.

The Future of Advanced Data Modes

The evolution of advanced data modes continues to shape the future of amateur radio. As technology progresses, new modes and enhancements to existing ones are developed, offering even greater capabilities. For example, enhancements to FT8 and the introduction of FT4, a faster variant, demonstrate the ongoing innovation in this field.

Amateur radio operators who embrace advanced data modes find themselves at the forefront of this technological evolution. These modes not only provide practical benefits for everyday operation but also offer a gateway to the cutting-edge aspects of radio communication. By exploring and utilizing advanced data modes, operators can enhance their communication abilities, contribute to the advancement of radio technology, and engage more deeply with the amateur radio community.

CHAPTER 12 – Community and Resources

Joining Ham Radio Clubs

Joining a ham radio club can be one of the most enriching experiences for any amateur radio operator. These clubs provide a sense of community, a wealth of knowledge, and numerous opportunities for learning and growth. Whether you are a newcomer to the hobby or a seasoned operator, becoming part of a ham radio club offers countless benefits and opens the door to a world of possibilities.

The Role of Ham Radio Clubs

Ham radio clubs serve as the backbone of the amateur radio community. They bring together individuals who share a common interest in radio communications, fostering an environment where members can share knowledge, collaborate on projects, and support each other's endeavors. These clubs often organize regular meetings, events, and activities that cater to the diverse interests of their members.

One of the primary roles of a ham radio club is education. Clubs frequently offer training sessions, workshops, and presentations on various topics related to amateur radio. These educational activities can range from basic licensing courses for beginners to advanced technical workshops on topics like antenna design, digital modes, and contesting. By participating in these educational activities, members can continuously expand their knowledge and skills.

Ham radio clubs also play a crucial role in promoting the hobby to the broader public. Many clubs participate in community events, demonstrations, and outreach programs to raise awareness about amateur radio. These activities not only help to attract new members but also highlight the importance of amateur radio in emergency communications and public service.

Types of Activities and Events

Clubs offer a wide variety of activities and events that cater to different aspects of the hobby. Here are some common types of activities you might encounter in a ham radio club:

Meetings: Regular club meetings provide a platform for members to discuss club business, share updates, and plan future activities. These meetings often include guest speakers who present on various topics of interest to the members.

Field Days: Field Day events are a staple of many ham radio clubs. During these events, members set up portable stations in outdoor locations and operate under simulated emergency conditions. Field Days provide an excellent opportunity to practice operating skills, test equipment, and enjoy the camaraderie of fellow operators.

Contests and Special Events: Many clubs organize or participate in radio contests and special event stations. These activities offer a chance to engage in friendly competition, make contacts with operators from around the world, and earn awards and recognition.

Workshops and Training Sessions: Clubs often hold workshops and training sessions on a wide range of topics. These can include everything from basic soldering and electronics skills to advanced topics like software-defined radio and digital communication modes.

Community Service: Ham radio clubs are often involved in community service activities, such as providing communications support for local events, participating in emergency preparedness drills, and assisting with disaster response efforts. These activities demonstrate the value of amateur radio to the broader community and provide members with a sense of purpose and fulfillment.

Finding the Right Club

With so many ham radio clubs available, finding the right one can seem daunting. Start by considering your interests and goals. Are you looking to improve your technical skills, participate in contests, or get involved in emergency communications? Identifying your priorities can help you narrow down your options.

Geographic location is another important factor. While many clubs have transitioned to online meetings and activities, having a local club can provide additional benefits, such as in-person events and hands-on workshops. Use resources like the ARRL (American Radio Relay League) website to find clubs in your area. Additionally, attending local hamfests and amateur radio conventions can be a great way to meet club members and learn about different clubs in your region.

Example Table: Key Activities Offered by Ham Radio Clubs

Activity Type	Description
Regular Meetings	Discuss club business, updates, and guest speaker presentations
Field Days	Portable station setup and operation under simulated conditions
Contests and Events	Participation in radio contests and special event stations
Workshops and Training	Sessions on various topics from basic to advanced
Community Service	Providing communications support for local events

Joining a ham radio club can significantly enhance your experience in the hobby. These clubs offer a supportive community, opportunities for continuous learning, and a platform to contribute to the broader community. Whether you're looking to build your technical skills, participate in contests, or engage in public service, being part of a ham radio club can help you achieve your goals and enrich your amateur radio journey.

Conclusion and BONUS

As we reach the end of this comprehensive guide to ham radio, it's essential to reflect on the journey we've undertaken and the vast world of amateur radio that lies ahead. Ham radio is more than just a hobby; it's a community, a technical challenge, and a platform for lifelong learning and exploration.

Throughout this book, we've covered the foundational aspects of getting started with ham radio, from licensing and regulations to the technical intricacies of radio equipment and operating practices. We've delved into the fascinating realm of radio technology fundamentals, explored the complexities of circuitry and design, and examined the vital elements of transmission lines and antennas. Each chapter has aimed to equip you with the knowledge and skills needed to become a proficient ham radio operator.

One of the most enriching aspects of ham radio is the sense of community it fosters. Joining a ham radio club, participating in contests, and engaging with other operators around the world are all integral parts of the experience. These interactions not only enhance your skills but also provide a network of support and camaraderie that can be both rewarding and inspiring.

The advanced topics we've explored, such as homebrewing equipment and utilizing software-defined radios, open up new dimensions for experimentation and personal growth. These areas allow you to push the boundaries of what's possible with amateur radio, contributing to the ongoing innovation and development within the field.

As you continue your journey in ham radio, remember that learning and improvement are ongoing processes. The landscape of amateur radio is always evolving, with new technologies, techniques, and opportunities emerging regularly. Staying curious, seeking out new challenges, and continuously building on your knowledge will keep the hobby engaging and fulfilling.

Finally, I encourage you to share your passion for ham radio with others. Whether through mentoring newcomers, participating in community service, or simply making contacts around the globe, your involvement helps to sustain and grow the ham radio community. By sharing your experiences and knowledge, you contribute to the rich tapestry of amateur radio and inspire the next generation of operators.

Thank you for embarking on this journey through the world of ham radio. May your future endeavors be filled with exciting discoveries, meaningful connections, and a deep sense of accomplishment. Keep exploring, keep learning, and most importantly, keep enjoying the wonderful world of amateur radio!

Download here the exam questions and answers, go to the link below or scan the QR code:

https://bit.ly/hamradiobonusquestions

Made in United States
Orlando, FL
14 October 2024